The ESSE

# STRENGTH OF MATERIALS
# and
# MECHANICS OF SOLIDS I

**Staff of Research and Education Association,
Dr. M. Fogiel, Director**

> This book covers the usual course outline of
> Strength of Materials and Mechanics of Solids I.
> For related topics, see *"THE ESSENTIALS OF
> STRENGTH OF MATERIALS and MECHANICS
> OF SOLIDS II"*.

 Research and Education Association
61 Ethel Road West
Piscataway, New Jersey 08854

# THE ESSENTIALS® OF STRENGTH OF MATERIALS & MECHANICS OF SOLIDS I

Printed in the United States of America

Library of Congress Catalog Card Number 89-62161

International Standard Book Number 0-87891-624-5

REVISED PRINTING, 1993

ESSENTIALS is a registered trademark of
Research and Education Association, Piscataway, New Jersey 08854

# WHAT "THE ESSENTIALS" WILL DO FOR YOU

This book is a review and study guide. It is comprehensive and it is concise.

It helps in preparing for exams, in doing homework, and remains a handy reference source at all times.

It condenses the vast amount of detail characteristic of the subject matter and summarizes the **essentials** of the field.

It will thus save hours of study and preparation time.

The book provides quick access to the important facts, principles, theorems, concepts, and equations in the field.

Materials needed for exams can be reviewed in summary form – eliminating the need to read and re-read many pages of textbook and class notes. The summaries will even tend to bring detail to mind that had been previously read or noted.

This "ESSENTIALS" book has been carefully prepared by educators and professionals and was subsequently reviewed by another group of editors to assure accuracy and maximum usefulness.

Dr. Max Fogiel
Program Director

# CONTENTS

**4     STRESS-STRAIN RELATIONS     28**

**5     CENTER OF GRAVITY, CENTROIDS AND MOMENT OF INERTIA     35**

## 6    STRESSES IN BEAMS    64

## 7    DESIGN OF BEAMS    73

# 8     DEFLECTION OF BEAMS     91

# CHAPTER 1

# AXIAL FORCE, SHEAR FORCE AND BENDING MOMENT

## 1.1    EQUILIBRIUM OF A SOLID BODY

### 1.1.1    SUMMATION OF THE FORCES ACTING ON A SOLID

$$\sum F_x = 0, \ \sum F_y = 0, \ \sum F_z = 0 \qquad (1\text{-}1)$$

$$\sum M_x = 0, \ \sum M_y = 0, \ \sum M_z = 0 \qquad (1\text{-}2)$$

### 1.1.2    THE FORCE VECTOR, $\overline{F}$

$$\overline{F} = F_x \overline{i} + F_y \overline{j} + F_z \overline{k} \qquad (1\text{-}3)$$

### 1.1.3    THE MOMENT VECTOR, $\overline{M}$

$$\overline{M} = M_x \overline{i} + M_y \overline{j} + M_z \overline{k} \qquad (1\text{-}4)$$

Where $F_x, F_y, F_z$ are components of the force, $F$, and $\overline{i}, \overline{j}, \overline{k}$ are

1

the unit vectors along the $x$, $y$ and $z$ axes, respectively.

Concurrently, $M_x$, $M_y$, $M_z$ are components of $M$ along the coordinate axes.

# 1.2 CONVENTIONS FOR SUPPORTS AND LOADINGS

### 1.2.1 LINK AND ROLLER SUPPORTS

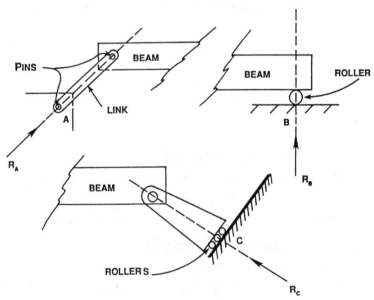

**FIGURE 1.1**–Link and Roller Supports resist only one force through a specific Line of Action (the Line of Action is indicated with dashes).

### 1.2.2 PINNED SUPPORTS

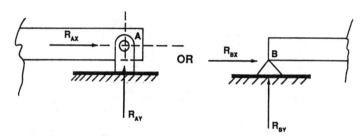

**FIGURE 1.2**–A Pinned Support is capable of resisting a force in any direction of the Plane. In general, it is resolved into two forces — Horizontal and Vertical.

### 1.2.3 FIXED SUPPORTS

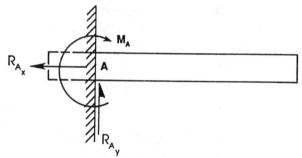

**FIGURE 1.3**–Fixed Supports can resist a force in any direction and are also capable of resisting a Couple or a Moment.

### 1.2.4 DISTRIBUTED LOADING

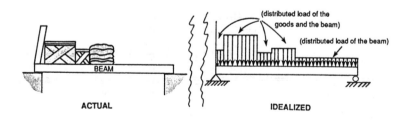

**FIGURE 1.4**–Simply supported Beam with uniformly distributed loads.

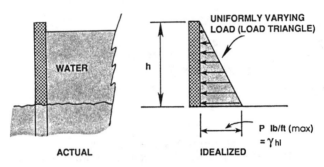

FIGURE 1.5–The force due to Hydrostatic Loading is uniformly varying. Here $\gamma$ is the unit weight of the liquid, and $P$ is the pressure (maximum) at the base.

**NOTE**: For a uniformly varying loading, the load may be resolved into a single load applied at 1/3 the distance from the base of the "Load Triangle" and equal to $F_T$, where

$$F_T = \frac{1}{2} Ph = \frac{1}{2} \gamma h^2 l \qquad (1\text{-}5)$$

# 1.3  AXIAL FORCE, SHEAR FORCE AND BENDING MOMENT DIAGRAMS

## 1.3.1    AXIAL FORCE IN BEAMS

The **axial force** is a force, $P$, acting through the beam's cross-sectional area, along the direction of the longitudinal axis.

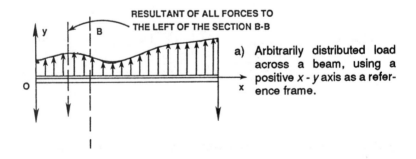

a) Arbitrarily distributed load across a beam, using a positive $x$ - $y$ axis as a reference frame.

4

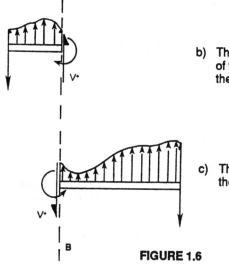

b) The shear, $V$, at one section of the beam is examined on the left of the section.

c) The shear is examined on the right of the section.

**FIGURE 1.6**

### 1.3.2 SHEAR

The **shear** or **shearing force** is the internal force, $V$, acting at right angles to the axis of the beam. It is numerically equal to the algebraic sum of all the external vertical forces acting on the isolated segment, but it is opposite in direction.

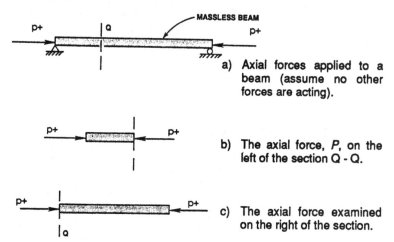

a) Axial forces applied to a beam (assume no other forces are acting).

b) The axial force, $P$, on the left of the section Q - Q.

c) The axial force examined on the right of the section.

**FIGURE 1.7**–Axial Force on a Beam.

5

### 1.3.3 BENDING MOMENT

The **bending moment** or the **internal resisting moment,** $M$, acts in the direction opposite to the external moment (or force couple) and it is equal in magnitude.

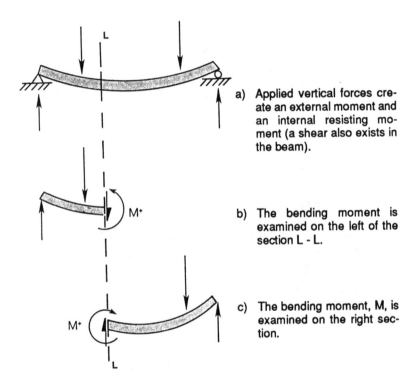

a) Applied vertical forces create an external moment and an internal resisting moment (a shear also exists in the beam).

b) The bending moment is examined on the left of the section L - L.

c) The bending moment, M, is examined on the right section.

FIGURE 1.8–Bending Moment in a Beam

## 1.3.4 AXIAL, SHEAR, AND BENDING MOMENT DIAGRAMS OF VARIOUS BEAMS AND BEAM LOADINGS (METHOD OF SECTIONS)

### 1.3.4.1 SIMPLY SUPPORTED BEAM

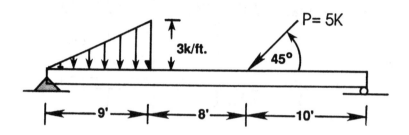

**FIGURE 1.9**–Simply Supported Beam with a Distributed Load and an Applied Force, *P*.

### FREE BODY DIAGRAM

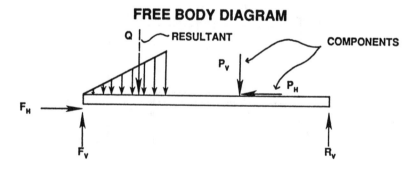

From equilibrium, we find the external forces:

$$F_H = 3.5k \quad F_V = 8k \quad P_V = 3.5k \quad P_H = 3.5k$$

$$Q = 9k \qquad R_V = 4.5$$

We can then analyze the beam using the method of sections.

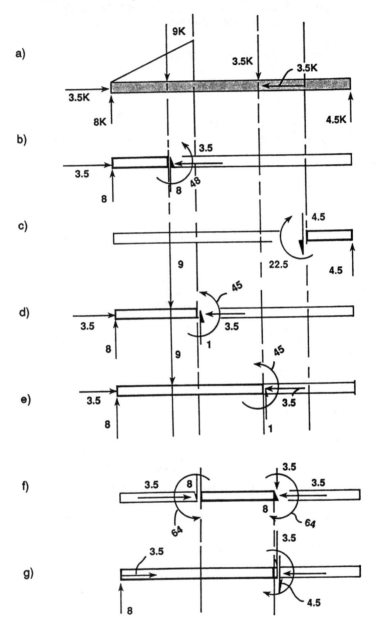

# 1.4 DIFFERENTIAL EQUATIONS OF EQUILIBRIUM

## 1.4.1 BASIC EQUATIONS:

$$\frac{dV}{dx} = -P \text{ and } \frac{dM}{dx} = -V \qquad (1\text{-}6)$$

$$\frac{d^2M}{dx^2} = P \qquad (1\text{-}7)$$

where  $P$ = Load
  $V$ = Shear
  $M$ = Bending Moment

## 1.4.2 SHEAR DIAGRAMS BY SUMMATION

Between any two definite sections of a beam, the change in shear is the negative of all the vertical forces included between these sections. If no force occurs between any two sections, no change in shear takes place.

If a concentrated force comes into the summation, a discontinuity in the value of the shear occurs (see Fig. 1.10).

The basic relation for shear, $V$:

$$V(x) = -\int_0^x p \, dx + C \qquad (1\text{-}8)$$

where $C$ = constant of integration.

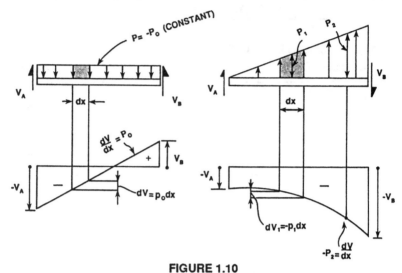

**FIGURE 1.10**

(a) Shear Diagram for a beam with a constant distributed load.

(b) Shear Diagram for a beam with a uniformly varying distributed load.

The change in shear depends entirely on whether the forces act up or down. If they act **upward**, the change is **negative.**

The **slope** of the shear diagram $\left(\dfrac{dv}{dx} = -p\right)$ determines the **rate of change in shear.** For a load of constant magnitude, the shear curve is a straight line. For a variable load, the shear curve is concave downward for an upward force and concave upward for a downward force.

The rate of change of the slope of the shear diagram equals the negative of the rate of change of the load.

### 1.4.3    MOMENT DIAGRAMS BY SUMMATION

If the ends of a beam are on rollers, the starting and terminal moments are zero. If the end is built in, the end moment is known from the reaction calculations (for statically determinate beams).

The change in moment in a given segment of a beam is equal to the negative of the area of the corresponding shear diagram.

The slope of the bending moment curve is determined by noting the corresponding magnitude and sign of the shear.

The basic relation for the moment, $M$:

$$M(x) = -\int_0^x V dx + C_1 \qquad (1\text{-}9)$$

where $C_1$ = constant of integration.

A constant shear yields a uniform rate of change in the bending moment, resulting in a straight line in the moment diagram. If no shear occurs along a certain portion of the beam, no change in moment occurs.

The maximum or minimum moment occurs at a point where the shear is zero, as the derivative of $M$ is then zero. This occurs at a point where the shear changes sign.

# CHAPTER 2

# STRESS

## 2.1    DEFINITION OF STRESS

### 2.1.1      GENERAL DEFINITION OF STRESS AT A POINT

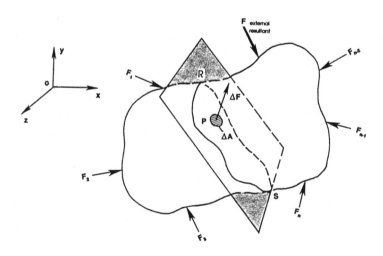

FIGURE 2.1–Arbitrary Solid Being Acted on by External Forces

12

Consider the elemental area $\Delta A$ at a point $P$ located in the cross-sectional plane, $RS$, with an internal resultant force $\Delta F$ acting on the area (See Fig. 2.1).

The stress, or force intensity, at $P$ is the limiting value of the ratio $\Delta F / \Delta A$ as the elemental area approaches zero. It is written as

$$\text{Stress at point } P = \left( \frac{\Delta F}{\Delta A} \right)_{\Delta A \to 0} \tag{2-1}$$

### 2.1.2 NORMAL STRESS, $\sigma$ (SIGMA)

The intensity of the forces perpendicular to the cross-sectional area is called the normal stress (tensile or compressive stress).

### 2.1.3 SHEARING STRESS, $\tau$ (TAU)

The intensity of the forces parallel to the plane of the cross-sectional area is called the shearing stress. (The components of stress are further defined in the following sections.)

### 2.1.4 MATHEMATICAL DEFINITION OF STRESS AT A POINT

$$\underbrace{\tau_{xx} = \lim_{\Delta A \to 0} \frac{\Delta F_x}{\Delta A}}_{\text{NORMAL STRESS}}, \; \underbrace{\tau_{xy} = \lim_{\Delta A \to 0} \frac{\Delta F_y}{\Delta A} \text{ and } \tau_{xz} = \lim_{\Delta A \to 0} \frac{\Delta F_z}{\Delta A}}_{\text{SHEARING STRESSES}}$$

(**NOTE**: The subscripts for the stresses indicate the plane in which they act. $\tau_{xx}$ is often written as $\sigma_x$, denoting a normal stress in the $x - x$ plane. It follows that $\tau_{yy} \equiv \sigma_y$ and $\tau_{zz} \equiv \sigma_z$.)

## 2.2 STRESS TENSOR

When considering the most general state of stress acting on an element, we isolate an infinitesimal cubic slice of the body. This cube may be called the stress tensor, and it may be represented in a matrix form as follows:

$$\begin{pmatrix} \tau_{xx} & \tau_{xy} & \tau_{xz} \\ \tau_{yx} & \tau_{yy} & \tau_{yz} \\ \tau_{zx} & \tau_{zy} & \tau_{zz} \end{pmatrix} \equiv \begin{pmatrix} \sigma_x & \tau_{xy} & \tau_{xz} \\ \tau_{yx} & \sigma_y & \tau_{yz} \\ \tau_{zx} & \tau_{zy} & \sigma_z \end{pmatrix} \qquad (2\text{-}2)$$

An examination of the stress symbols shown in Figure 2.2 shows that there are three normal stresses

$$\tau_{xx} \equiv \sigma_x , \ \tau_{yy} \equiv \sigma_y , \ \tau_{zz} \equiv \sigma_z ,$$

and six shearing stresses

$$\tau_{xy}, \tau_{xz}, \ \tau_{yx}, \tau_{yz}, \tau_{zx}, \tau_{zy} .$$

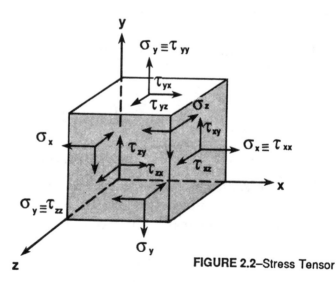

FIGURE 2.2–Stress Tensor

14

# 2.3 STRESS IN AXIALLY LOADED MEMBERS

### 2.3.1 NORMAL STRESS (PERPENDICULAR TO THE CUT)

$$\sigma = \frac{P}{A}$$

(2-3)

$\sigma =$ Normal stress uniformly distributed over the cross-sectional area

$P =$ Force

$A =$ Area

### 2.3.2 BEARING STRESS

If the resultant of the applied forces coincides with the centroid of the contact area between the two bodies, the normal stress, $\sigma$, is called a **bearing stress**. One body supported by another creates a bearing stress.

### 2.3.3 AVERAGE SHEARING STRESS (PARALLEL TO THE CUT)

$$\tau = \frac{P}{A}$$

(2-4)

$\tau =$ Shearing stress uniformly distributed across the cross-sectional area

$P =$ Force

$A =$ Area

## 2.4    DESIGN OF AXIALLY LOADED MEMBERS AND PINS

### 2.4.1    REQUIRED AREA OF A MEMBER

$$A = \frac{P}{\sigma_a} \qquad \text{(2-5)}$$

$A =$  Required Area
$P =$  Axial Force
$\sigma_a =$  Allowable Stress *

\* The allowable stress is determined by dividing the normal stress by a given safety factor,

$$\sigma_a = \frac{\sigma}{S.F.} \qquad \text{(2-6)}$$

### 2.4.2    EXAMPLE

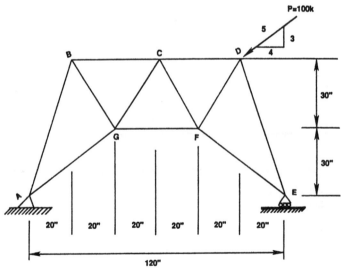

**FIGURE 2.3**–Example Problem

16

Choose members *BC* and *CF* in the truss of Figure 2.3 to carry a force *P* of 100 Kips, acting as shown. The allowable tensile stress is 20 Kpsi. Find the cross-sectional areas of the members.

For overall equilibrium,

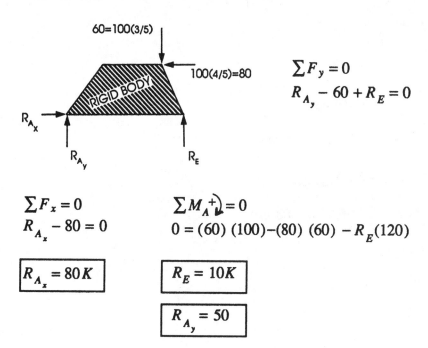

$$60 = 100(3/5)$$

$$100(4/5) = 80$$

$$\sum F_y = 0$$
$$R_{A_y} - 60 + R_E = 0$$

$\sum F_x = 0$
$R_{A_x} - 80 = 0$

$\sum M_A^+ = 0$
$0 = (60)(100) - (80)(60) - R_E(120)$

$$\boxed{R_{A_x} = 80\,K}$$

$$\boxed{R_E = 10\,K}$$

$$\boxed{R_{A_y} = 50}$$

Then, using the method of sections, we solve for *BC* and *CF*.

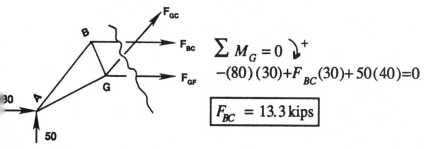

$\sum M_G = 0$
$-(80)(30) + F_{BC}(30) + 50(40) = 0$

$$\boxed{F_{BC} = 13.3\,\text{kips}}$$

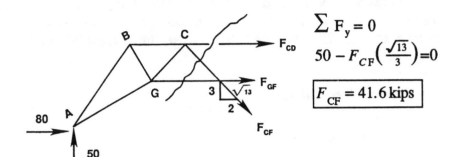

$$\sum F_y = 0$$

$$50 - F_{CF}\left(\frac{\sqrt{13}}{3}\right) = 0$$

$$\boxed{F_{CF} = 41.6 \text{ kips}}$$

$A_{BC} = F_{BC}/\sigma_{\text{allowable}}$

$A_{BC} = 13.3/20 = .667 \text{ in}^2$

$A_{CF} = F_{CF}/\sigma_{\text{allowable}}$

$A_{CF} = 41.6/20 = 2.08 \text{ in}^2$

# CHAPTER 3

# STRAIN

## 3.1 DEFINITION OF STRAIN

### 3.1.1 GENERAL DEFINITION OF STRAIN

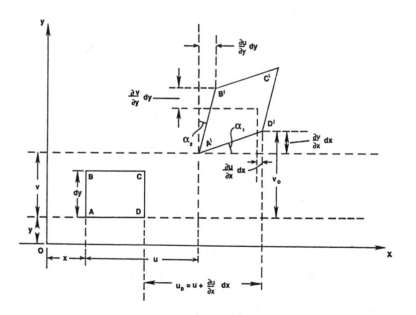

FIGURE 3.1–Rectangular Element Before and After Deformation.

Consider a small rectangular element *ABCD* in a homogeneous, isotropic body, with sides *dx* and *dy* in the plane before deformation. External forces are applied to the body, and the element is displaced to a final position *A'B'C'D'* (See Figure 3.1).

The displacement consists of two basic geometric deformations of the element:

- A change in length of an initial straight line in a certain direction

- A change in the value of the given angle.

These deformations are classified, respectively, as the longitudinal strain and the shear strain.

### 3.1.2   LONGITUDINAL STRAIN, ε (EPSILON)

The ratio of the change in length, $\Delta l$, to the initial length, $l$, of a straight-line element.

### 3.1.3   SHEAR STRAIN, γ (GAMMA)

The change in value of the initial right angle between the straight-line elements *AB* and *AD* (See Figure 3.1).

### 3.1.4   MATHEMATICAL DEFINITIONS OF STRAIN

### 3.1.4.1   LONGITUDINAL STRAIN, ε

$$\varepsilon_x = \frac{\partial u}{\partial x} , \; \varepsilon_y = \frac{\partial v}{\partial y} , \; \varepsilon_z = \frac{\partial w}{\partial z} \tag{3-1}$$

where *u*, *v*, and *w* are the three displacement components, at a point of the body, occurring, respectively, in the *x*, *y*, and *z* directions of the coordinate axes.

$$\gamma_{xy} = \gamma_{yx} = \frac{\partial u}{\partial x} + \frac{\partial u}{\partial y} \qquad (3\text{-}2)$$

$$\gamma_{xz} = \gamma_{zx} = \frac{\partial w}{\partial x} + \frac{\partial u}{\partial z} \qquad (3\text{-}3)$$

$$\gamma_{yz} = \gamma_{zy} = \frac{\partial w}{\partial y} + \frac{\partial v}{\partial z} \qquad (3\text{-}4)$$

where u, v, and w are the three displacement components, at a point of the body, occurring, respectively, in the $x$, $y$, and $z$ directions of the coordinate axes.

## 3.2    STRAIN TENSOR

The general state of strain, like that of stress, may be represented in a matrix form as follows:

$$\begin{pmatrix} \varepsilon_x & \dfrac{\gamma_{xy}}{2} & \dfrac{\gamma_{xz}}{2} \\[2ex] \dfrac{\gamma_{yx}}{2} & \varepsilon_y & \dfrac{\gamma_{yz}}{2} \\[2ex] \dfrac{\gamma_{zx}}{2} & \dfrac{\gamma_{zy}}{2} & \varepsilon_z \end{pmatrix} \equiv \begin{pmatrix} \varepsilon_{xx} & \varepsilon_{xy} & \varepsilon_{xz} \\[1ex] \varepsilon_{yx} & \varepsilon_{yy} & \varepsilon_{yz} \\[1ex] \varepsilon_{zx} & \varepsilon_{zy} & \varepsilon_{zz} \end{pmatrix} \qquad (3\text{-}5)$$

## 3.3    HOOKE'S LAW FOR ISOTROPIC MATERIALS

The linear relationships between the six components of stress and the six components of strain may be generalized and simplified, as stated in Hooke's Laws for homogeneous, iso-tropic materials.

### 3.3.1 GENERAL EQUATIONS

$$\varepsilon_x = \frac{\sigma_x}{E} - \nu\frac{\sigma_y}{E} - \nu\frac{\sigma_z}{E} \qquad (3\text{-}6)$$

$$\varepsilon_y = -\frac{\nu\sigma_x}{E} + \frac{\sigma_y}{E} - \nu\frac{\sigma_z}{E} \qquad (3\text{-}7)$$

$$\varepsilon_z = -\frac{\nu\sigma_x}{E} - \frac{\nu\sigma_y}{E} + \frac{\sigma_z}{E} \qquad (3\text{-}8)$$

and

$$\gamma_{xy} = \frac{\tau_{xy}}{G} \;,\; \gamma_{yz} = \frac{\tau_{yz}}{G} \;,\; \gamma_{zx} = \frac{\tau_{zx}}{G} \qquad (3\text{-}9)$$

where   $E$ = a constant called Young's Modulus or the Modulus of Elasticity

$G$ = a constant called the Shearing Modulus of Elasticity or the Modulus of Rigidity

$\nu$ = Poisson's Ratio (a strain relation)

The equation relating all three constants is:

$$G = \frac{E}{2(1 + \nu)} \qquad (3\text{-}10)$$

# 3.4    POISSON'S RATIO

$$\nu = -\frac{\varepsilon_y}{\varepsilon_x} = -\frac{\varepsilon_z}{\varepsilon_x}$$

(3-11)

$\nu$ =   Poisson's Ratio
$\varepsilon_y$ =   Lateral Strain
$\varepsilon_z$ =   Lateral Strain
$\varepsilon_x$ =   Axial  Strain

# 3.5    THERMAL STRAIN

Strains due to the temperature change

$$\varepsilon_x = \varepsilon_y = \varepsilon_z = \alpha\delta T$$

(3-12)

$\varepsilon_x, \varepsilon_y, \varepsilon_z$    = Thermal strains
$\alpha$            = Coefficient of linear thermal expansion
$\delta T$            = Change in temperature

**NOTE:** For an increase in temperature, $\delta T$ is taken positive. For example,

$$\varepsilon_z = -\frac{\nu\sigma_x}{E} - \frac{\nu\sigma_y}{E} + \frac{\sigma_z}{E} + \alpha\delta T$$

(3-13)

# 3.6    ELASTIC STRAIN ENERGY

The elastic strain energy is the internal work done in a body due to externally applied forces.

**3.6.1**

$$dU = \frac{1}{2}\sigma_x \varepsilon_x dV \qquad \text{(3-14)}$$

$dU$ = The elemental, internal elastic strain energy for uniaxial stress

$\sigma_x$ = Normal stress

$\varepsilon_x$ = Linear (longitudinal) strain

$dV$ = Volume of the element

The strain energy density is the strain energy stored in an elastic body per unit volume of the material.

**3.6.2**

$$\frac{dU}{dV} = U_0 = \frac{\sigma_x \varepsilon_x}{2} \qquad \text{(3-15)}$$

where $U_0$ = strain-energy density for uniaxial stress.

**3.6.3**

$$dU_s = \frac{1}{2}\tau_{xy}\gamma_{xy}dV \qquad \text{(3-16)}$$

$dU_s$ = The elemental, internal elastic strain energy for shearing stresses

$\tau_{xy}$ = Shear stress

$\gamma_{xy}$ = Displacement due to deformation

$dV$ = Volume of the element

**3.6.4**

$$\left(\frac{dU}{dV}\right)_s = (U_0)_s = \frac{\tau_{xy}\gamma_{xy}}{2} \qquad \text{(3-17)}$$

where $(U_0)_s$ = Strain-energy density for shearing stresses.

**3.6.5** General equation for strain-energy density having multiaxial states of stress.

$$U_0 = \frac{1}{2E}\left(\sigma_x^2 + \sigma_y^2 + \sigma_z^2\right) - \frac{v}{E}\left(\sigma_x\sigma_y + \sigma_y\sigma_z + \sigma_z\sigma_x\right)$$
$$+ \frac{1}{2G}\left(\tau_{xy}^2 + \tau_{yz}^2 + \tau_{zx}^2\right) \tag{3-18}$$

# 3.7 DEFLECTION OF AXIALLY LOADED MEMBERS

With applied forces only, the deflection of a beam, axially loaded, may be represented as:

$$u = \sum \frac{PL}{AE} \tag{3-19}$$

$u$ = Total deflection (displacement)
$P$ = Applied force
$L$ = Length over which the force is acting
$A$ = Cross-sectional area of the member
$E$ = Modulus of elasticity

For a fixed-supported rod free at one end, the deflection at the free end caused by an applied force is

$$u(L) = \frac{PL}{AE} \tag{3-20}$$

where $u(L)$ is the deflection as a function of the length of the rod.

The deflection caused by the applied force **and** the weight of the rod may be written as:

$$|u| = \frac{PL}{AE} + \frac{WL}{2AE} = \frac{\left[P + (\frac{W}{2})\right]L}{AE}$$

(3-21)

where $W$ is the weight of the rod.

### STIFFNESS

$$k = \frac{AE}{L}$$

(3-22)

$k$ = Stiffness influence coefficient
$A$ = Cross-sectional area
$E$ = Modulus of elasticity
$L$ = Length of the rod

(The reciprocal of $k$ defines the flexibility coefficient $f = k^{-1}$).

# 3.8   STRESS CONCENTRATIONS

When a rigid body is undergoing stress, no matter how irregular the stress distribution is at a given section, the sum of the stress must be equal to the applied force. However, calculations for peak (maximum) stresses may be computed by the following equation:

$$\sigma_{max} = k\frac{P}{A}$$

(3-23)

where $K$ is a calculate constant varying from material to

26

material. $K$ may also be calculated knowing the maximum stress, $\sigma_{max}$, and the average stress, $\sigma_{avg}$.

$$K = \frac{\sigma_{max}}{\sigma_{avg}}$$

# CHAPTER 4

# STRESS-STRAIN RELATIONS

## 4.1    MATERIAL PROPERTIES

### 4.1.1    DEFINITIONS

**Ductile materials** are capable of withstanding large strains. The converse applies to **brittle materials.**

**True stress** is obtained by dividing the applied force by the corresponding actual area of a specimen at the same instant. This differs from the **conventional** or **engineering stresses,** which are computed on the basis of the original area of a specimen.

The **proportional limit** of a material may be roughly defined as the point on the stress-strain diagram where the equation for the slope of the $\sigma - \varepsilon$ curve, due to an applied load, is no longer accurately approximated as being linear.

The slope of the line from 0 (the unloaded starting point of the diagram) to $A$ (the proportional limit) is the elastic modulus, $E$. This is also a measure of stiffness of the material, due to an imposed load.

The highest point in the curve on a $\sigma - \varepsilon$ diagram is said to correspond with the **ultimate strength** of a material.

The **yield point** of a material exists only in ductile substances. It is the point where a large amount of deformation occurs due to an essentially constant stress.

Often, the yield point is nearly indistinguishable from the proportional limit, and for experimental purposes, the **offset method** is used to make this point more distinct. (See Figure 4.1.) Here, a line offset an aribtrary amount (0.2 percent) of strain is drawn parallel to the straight-line portion of the initial stress-strain diagram. The point $Q$ is then taken to be the yield point of the material at 0.2 percent offset.

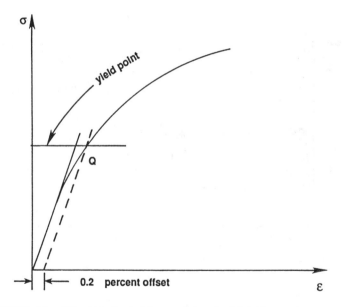

FIGURE 4.1—Offset Method of Determining the Yield Point of a Material.

The **critical point** of a material is the point (approximately) where a certain substance ruptures, under specific experimental conditions, due to an applied tensile loading.

29

### 4.1.2     ELASTICITY

An elastic material is one that is able to **completely** regain its original dimensions upon the removal of the applied forces. This implies that no permanent deformation occurs under any given conditions.

The point at which **permanent set** or **permanent deformation** occurs in a material is called the **elastic limit**. Non-elastic reactions are called **plastic**.

# 4.2     STRESS-STRAIN DIAGRAMS

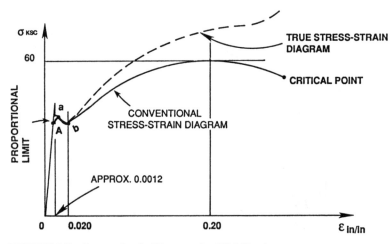

**FIGURE 4.2**—Stress-Strain Diagram for Mild Steel.

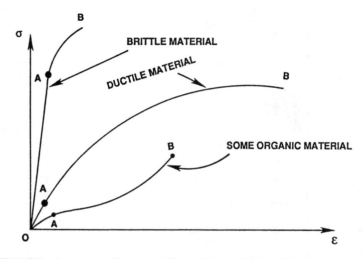

**FIGURE 4.3—**$\sigma - \varepsilon$ Diagram of Three Types of Materials.

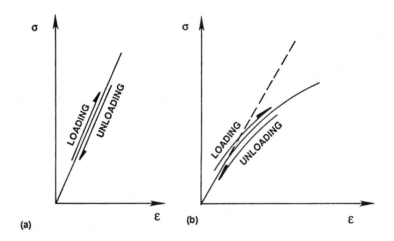

**FIGURE 4.4—(a)** Linearly Elastic Materials and **(b)** Nonlinearly Elastic Materials.

# 4.3 TABLE OF STRENGTH OF MATERIALS

## METALS AND ALLOYS

| Material | Stress in Kips per Square Inch | | | | | Modulus of Elasticity Pounds per Sq. In. | Elongation Percent |
| --- | --- | --- | --- | --- | --- | --- | --- |
| | Tension Ultimate | Elastic Limit | Compression Ultimate | Bending Ultimate | Shearing Ultimate | | |
| Aluminum, bars, sheets | 24-28 | 12-14 | | | | | |
| " wire, annealed | 20-35 | 14 | | | | | |
| Brass, 50% Zn | 31 | 17.9 | 117 | 33.5 | | — | 5.0 |
| " cast, common | 18-24 | 6 | 30 | 20 | 36 | 9,000,000 | |
| " wire, hard | 80 | 16 | | | | | |
| " annealed | 50 | 40 | | | | | |
| Bronze, aluminum 5 to 7½% | 75 | 40 | 120 | — | — | 14,000,000 | |
| " Tobin, cast 38% Zn | 66 | | | | | | |
| " rolled 1½% Sn | 80 | 40 | | | | 14,500,000 | |
| " c. " ⅓% Pb | 100 | | | | | | |
| Copper, plates, rods, bolts | 32-35 | 10 | 32 | | | | |
| Iron, cast, gray | 18-24 | 15-20 | | 25-33 | | | |
| " malleable | 27-35 | 26 | 46 | 30 | 40 | | |
| " wrought, shapes | 48 | | Tensile | Tensile | ⅚ Tens. | 28,000,000 | |
| Steel, plates for cold pressing | 48-58 | ½ Tens. | Tensile | Tensile | ¾ Tens. | 29,000,000 | *1,500,000 ÷ Tensile Strength |
| " cars | 50-65 | ½ Tens. | Tensile | Tensile | ¾ Tens. | 29,000,000 | |

| | | 1/2 Tens. | | | 3/4 Tens. | | *1,500,000 Tensile Strength |
|---|---|---|---|---|---|---|---|
| " locos., stat. boilers | 55-65 | 1/2 Tens. | Tensile | Tensile | 3/4 Tens. | 29,000,000 | |
| " bridges and bldgs., ships | 60-72 | 33 | Tensile | Tensile | 3/4 Tens. | 29,000,000 | |
| " structural silicon | 80-95 | 45 | Tensile | Tensile | 3/4 Tens. | 29,000,000 | |
| " struc. nickel (3.25% Ni) | 85-100 | 50 | Tensile | Tensile | 3/4 Tens. | 29,000,000 | |
| Steel, rivet, boiler | 45-55 | 1/2 Tens. | Tensile | Tensile | 3/4 Tens. | 29,000,000 | |
| " " br.,bldg.,loco.,cars | 52-62 | 28 | Tensile | Tensile | 3/4 Tens. | 29,000,000 | |
| " " ships | 55-65 | 30 | Tensile | Tensile | 3/4 Tens. | 29,000,000 | |
| " " high-tensile | 70-85 | 38 | Tensile | Tensile | 3/4 Tens. | 29,000,000 | |
| Steel, cast, soft | 60 | 27 | Tensile | Tensile | 3/4 Tens. | 29,000,000 | |
| " medium | 70 | 31.5 | Tensile | Tensile | 3/4 Tens. | 29,000,000 | |
| " hard | 80 | 36 | Tensile | Tensile | 3/4 Tens. | 29,000,000 | |
| Steel wire, unannealed | 120 | 60 | | | | 29,000,000 | †24 |
| " " annealed | 80 | 40 | | | | 29,000,000 | †20 |
| " " bridge cable | 215 | 95 | | | | 29,000,000 | †17 |

* 8" gage length
† 2" gage length

33

## BUILDING MATERIALS

| Material | Average Ultimate Stress Pounds per Square Inch | | | Safe Working Stress Pounds per Square Inch | | | Modulus of Elasticity Pounds Per Sq. In. |
|---|---|---|---|---|---|---|---|
| | Compression | Tension | Bending | Compression | Bearing | Shearing | |
| Masonry, granite | -- | -- | -- | 420 | 600 | | |
| " limestone, bluestone | -- | -- | -- | 350 | 500 | | |
| " sandstone | -- | -- | -- | 280 | 400 | | |
| " rubble | -- | -- | -- | 140 | 250 | | |
| " brick, common | 10,000 | 200 | 600 | | | | |
| Ropes, cast steel hoisting | -- | 80,000 | | | | | |
| " standing, derrick | -- | 70,000 | | | | | |
| " manila | -- | 8,000 | | | | | |
| Stone, bluestone | 12,000 | 1,200 | 2,500 | 1,200 | 1,200 | 200 | 7,000,000 |
| " granite, gneiss | 12,000 | 1,200 | 1,600 | 1,200 | 1,200 | 200 | 7,000,000 |
| " limestone, marble | 8,000 | 800 | 1,500 | 800 | 800 | 150 | 7,000,000 |
| " sandstone | 5,000 | 150 | 1,200 | 500 | 500 | 150 | 3,000,000 |
| " slate | 10,000 | 3,000 | 5,000 | 1,000 | 1,000 | 175 | 14,000,000 |

# CHAPTER 5

# CENTER OF GRAVITY, CENTROIDS AND MOMENT OF INERTIA

## 5.1 CENTER OF GRAVITY

For a System of Particles:

$$\bar{x} = \frac{\sum\limits_{i=1}^{n} x_i w_i}{\sum\limits_{i=1}^{n} w_i} \ , \ \bar{y} = \frac{\sum\limits_{i=1}^{n} y_i w_i}{\sum\limits_{i=1}^{n} w_i} \ , \ \bar{z} = \frac{\sum\limits_{i=1}^{n} z_i w_i}{\sum\limits_{i=1}^{n} w_i} \tag{5-1}$$

$\bar{x}, \bar{y}, \bar{z}$      are the coordinates of the center of gravity of the system.

$w_i$      is the weight of the $i$th particle.

$x_i, y_i, z_i$      are the algebraic distances to the $i$th particle from the origin.

## TWO-DIMENSIONAL CONDITIONS (AREAS AND LINES)

$$W = \int dw$$

$$\bar{x} = \frac{\int xdw}{W}$$

$$\bar{y} = \frac{\int ydw}{W}$$

(5-2)

$W$ is the sum of the magnitudes of the elementary weights. In the case of a line, the center of gravity generally does not lie on the line.

## THREE-DIMENSIONAL CONDITIONS (VOLUMES)

$$\bar{x} = \frac{\int xdw}{W} \; , \; \bar{y} = \frac{\int ydw}{W} \; , \; \bar{z} = \frac{\int zdw}{W}$$

(5-3)

If the body is made of homogeneous material of specific weight $\gamma$, then

Weight of the element $(dw) = \gamma \, dV$ where $dV$
= volume of the element and

$$\bar{x} = \frac{\int xdv}{V} \; , \; \bar{y} = \frac{\int ydv}{V} \; , \; \bar{z} = \frac{\int zdv}{V}$$

(5-4)

$\bar{x}, \bar{y}, \bar{z}$      are the coordinates of the centroid of the volume $V$ of the body.

$\int xdv$      is called the first moment of the volume with respect to the $yz$ plane.

36

# 5.2 CENTROIDS

When the calculation depends on the geometry of the body only, the point $(\bar{x}, \bar{y}, \bar{z})$ is called the centroid of the body.

For an area $A$ of a homogeneous plate:

$$\bar{x} = \frac{\int x \, dA}{A}$$

$$\bar{y} = \frac{\int y \, dA}{A}$$

(5-5)

The above equations define the coordinates $\bar{x}$ and $\bar{y}$ of the center of gravity. This point is also known as centroid of the area $A$.

For a homogeneous wire of length $L$:

$$\bar{x} = \frac{\int x \, dL}{L}$$

$$\bar{y} = \frac{\int y \, dL}{L}$$

(5-6)

**NOTE:** If the body is nonhomogeneous, the centroid and the center of gravity will not coincide.

### 5.2.1   COMPOSITE BODIES

A composite body may be divided into parts with common shapes to determine its center of gravity.

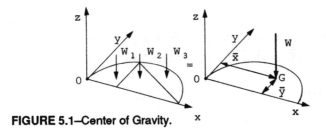

**FIGURE 5.1—Center of Gravity.**

By equating moments:

$$\bar{x} = \frac{\bar{x}_1 W_1 + \bar{x}_2 W_2 + \ldots + \bar{x}_n W_n}{W_1 + W_2 + \ldots + W_n}$$

$$\bar{y} = \frac{\bar{y}_1 W_1 + \bar{y}_2 W_2 + \ldots + \bar{y}_n W_n}{W_1 + W_2 + \ldots + W_n}$$

(5-7)

To find the centroid, equate moment of areas:

$$\bar{x} = \frac{\bar{x}_1 A_1 + \bar{x}_2 A_2 + \ldots + \bar{x}_n A_n}{A_1 + A_2 + \ldots + A_n}$$

$$\bar{y} = \frac{\bar{y}_1 A_1 + \bar{y}_2 A_2 + \ldots + \bar{y}_n A_n}{A_1 + A_2 + \ldots + A_n}$$

(5-8)

## LINES

For composite lines, divide it into simpler segments and add the moments of each segment.

## VOLUMES

A body can be divided into common shapes of revolution.

Center of Gravity:

$$\bar{x}\sum_{i=1}^{n} W_i = \sum_{i=1}^{n} \bar{x}_i W_i$$

$$\bar{y}\sum_{i=1}^{n} W_i = \sum_{i=1}^{n} \bar{y}_i W_i \qquad (5\text{-}9)$$

$$\bar{z}\sum_{i=1}^{n} W_i = \sum_{i=1}^{n} \bar{z}_i W_i$$

If the body is homogeneous, the center of gravity coincides with the centroid:

$$\bar{x}\sum_{i=1}^{n} V_i = \sum_{i=1}^{n} \bar{x}_i V_i$$

$$\bar{y}\sum_{i=1}^{n} V_i = \sum_{i=1}^{n} \bar{y}_i V_i \qquad (5\text{-}10)$$

$$\bar{z}\sum_{i=1}^{n} V_i = \sum_{i=1}^{n} \bar{z}_i V_i$$

## 5.2.2    CENTROIDS BY INTEGRATION

$$\bar{x} A = \int \bar{x}_e \, dA$$

$$\bar{y} A = \int \bar{y}_e \, dA \qquad (5\text{-}11)$$

$x_e, y_e$    are the coordinates of the centroid of the element $dA$.

$x_e, y_e$    should be expressed in terms of the coordinates of a point located on the curve bounding the area.

$dA$    should be expressed in terms of the coordinates of the point and their differentials.

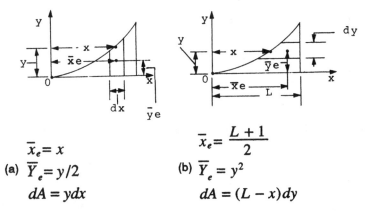

$$\overline{x}_e = x$$
(a) $\overline{Y}_e = y/2$
$$dA = y\,dx$$

$$\overline{x}_e = \frac{L + 1}{2}$$
(b) $\overline{Y}_e = y^2$
$$dA = (L - x)\,dy$$

**FIGURE 5.2–Centroids by Integration.**

## LINES

$$\overline{x}L = \int x\,dL$$

$$\overline{y}L = \int y\,dL$$

(5-12)

$dL$    should be expressed by one of the following:

$$dL = \sqrt{\left(\frac{dy}{dx}\right)^2 + 1}\ \ dx$$

$$dL = \sqrt{\left(\frac{dx}{dy}\right)^2 + 1}\ \ dy$$

$$dL = \sqrt{r^2 + \left(\frac{dr}{d\theta}\right)^2}\ d\theta$$

The above equations for $dL$, are chosen depending on the type of equation used to define the line.

**VOLUME**

$$\overline{x}V = \int \overline{x}_e\ dV$$

$$\overline{y}V = \int \overline{y}_e\ dV$$

$$\overline{z}V = \int \overline{z}_e\ dV$$

(5-13)

For bodies with two planes of symmetry:

$$\overline{y} = \overline{z} = 0$$

and

$$\overline{x}V = \int \overline{x}_e dV$$

# 5.3 THEOREMS OF PAPPUS-GULDINUS

Surface of Revolution – Generated by revolving a curve about a fixed axis.

**THEOREM I**

The area of a surface revolution is equal to the product of the length of the generating curve and the distance traveled by the centroid of the curve while the surface is being generated.

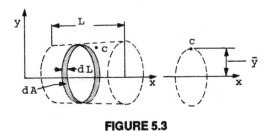

**FIGURE 5.3**

When a differential length $dL$ of line $L$ is revolved about $x$-axis, a ring is generated having surface area $dA = 2\pi y dL$. Therefore

$$A = 2\pi \int_L y \; dL = 2\pi \bar{y} L \qquad \text{(5-14)}$$

## THEOREM II

The volume of a surface of revolution is equal to the generating area times the distance traveled by the centroid of the area in generating the volume.

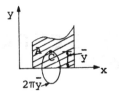

Volume of Revolution – Generated by revolving an area about a fixed axis. When a differential area $dA$ is revolved about $x$-axis, it generated a ring of volume $dV = 2\pi y dA$. Therefore

42

$$v = 2\pi \int_A y \; dA = 2\pi \overline{y} A \qquad \text{(5-15)}$$

## 5.4 MOMENT OF INERTIA

**AREA**

General Formula:

$$I = \int_A s^2 \, dA \qquad \text{(5-16)}$$

$s$ = Perpendicular distance from the axis to the area element (Figure 5.4).

In component form:

$$I_x = \int y^2 \, dA$$

$$I_y = \int x^2 \, dA \qquad \text{(5-17)}$$

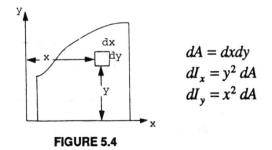

$$dA = dx\,dy$$
$$dI_x = y^2 \, dA$$
$$dI_y = x^2 \, dA$$

**FIGURE 5.4**

For a Rectangular Area:

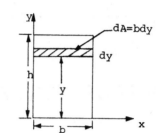

**FIGURE 5.5**

$$I_x = \int_0^h by^2 dy = \tfrac{1}{3} bh^3$$

(5-18)

**NOTE**: This is the moment of inertia with respect to an axis passing through the base of the rectangle.

Moments of Inertia of Masses

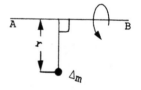

**FIGURE 5.6**

$$I = \int r^2 dm$$

(5-19)

In component form:

$$I_x = \int (y^2 + z^2)\, dm$$

$$I_y = \int (z^2 + x^2)\, dm$$

(5-20)

$$I_z = \int (x^2 + y^2)\, dm$$

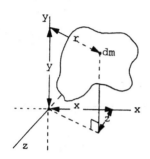

**FIGURE 5.7**

## 5.4.1    POLAR MOMENT OF INERTIA

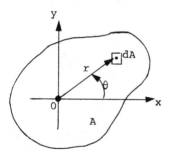

**FIGURE 5.8**

Polar Moment of Inertia:

$$J_0 = \int r^2 dA$$

(5-21)

In terms of rectangular moments of inertia:

$$J_0 = I_x + I_y \qquad \text{(5-22)}$$

## 5.4.2 RADIUS OF GYRATION

**AREAS**

$$I_x = k_x^2 A \ , \ I_y = k_y^2 A$$

Rectangular component form:

$$k_x = \sqrt{\frac{I_x}{A}}$$

$$k_y = \sqrt{\frac{I_y}{A}} \qquad \text{(5-23)}$$

**POLAR FORM**

$$k_0 = \sqrt{\frac{J_0}{A}} \qquad \text{(5-24)}$$

Relation between rectangular component form and polar form:

$$k_0^2 = k_x^2 + k_y^2 \qquad \text{(5-25)}$$

**MASSES**

$$I = k^2 m$$

$$k = \sqrt{\frac{I}{m}} \qquad \text{(5-26)}$$

### 5.4.3 PERPENDICULAR-AXIS THEOREM

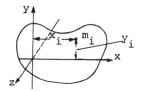

**FIGURE 5.9**

For a thin plane lamina:

$$I_z = \sum_i m_i x_i^2 + \sum_i m_i y_i^2 = I_x + I_y \; ; \; i = 1, 2, \ldots \qquad \text{(5-27)}$$

**STATEMENT**

The moment of inertia of any plane lamina about an axis normal to the plane of the lamina is equal to the sum of the moments of inertia about any two mutually perpendicular axes passing through the given axis and lying in the plane of the lamina.

### 5.4.4 PARALLEL-AXIS THEOREM

**AREAS**

The theorem states that the moment of inertia of an area about a given axis is equal to the sum of the moment of inertia parallel to the given axis (and passing through the centroid of the area) and the product of the area and the square of the distance between the two parallel axes.

$$I = I_c + Ad^2$$

(5-28)

$I_c$ = Moment of inertia about an axis through the centroid
$d$ = Distance Between the Axes
$A$ = Area

In terms of the radius of gyration:

$$k_0^2 = k_c^2 + d^2$$

(5-29)

In polar form:

$$J_0 = J_c + Ad^2$$

(5-30)

$J_c$ = Polar Moment of Inertia about the Centroid

**MASSES**

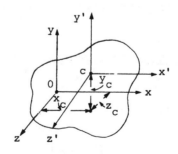

**FIGURE 5.10**

48

$$I = I_c + md^2 \qquad \text{(5-31)}$$

In the component form:

$$I_x = I_{x_c} + m(y_c^2 + z_c^2)$$

$$I_y = I_{y_c} + m(z_c^2 + x_c^2) \qquad \text{(5-32)}$$

$$I_z = I_{z_c} + m(x_c^2 + y_c^2)$$

### 5.4.5 COMPOSITE AREAS

A composite area consists of several regular shapes of bodies as components. The moments of inertia of these component shapes about a given axis may be computed separately and then added to give the moment of inertia for the entire area (or body).

The moment of inertia of a region with a hole equals the difference between the moment of inertia of the complete area (neglecting the hole) and the moment of inertia of the hole.

### EXAMPLE

$C$ is the centroid,

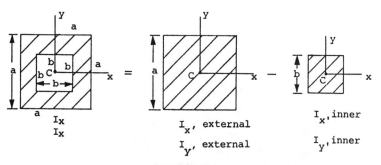

**FIGURE 5.11**

49

# 5.5    PRODUCT OF INERTIA

**AREAS**

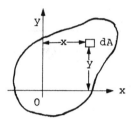

**FIGURE 5.12**

$$I_{xy} = \int_A xy\, dA$$

(5-33)

unit: (Length)$^4$

**NOTE**: $I_{xy}$ is zero when $x$, $y$, or both $x$ and $y$ axes are axes of symmetry for the region.

Parallel-Axis Theorem:

$$I_{xy} = I_{x'y'} + x_c y_c A$$

(5-34)

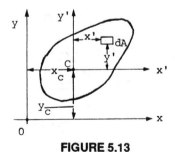

**FIGURE 5.13**

$I_{x'y'}$ = Moment of inertia in the centroidal axis system.

$x_c, y_c$ = Coordinates of the centroid in the '*xoy*' axis system.

## MASS PRODUCTS OF INERTIA

Component Equations:

$$I_{xy} = \int xy\,dm$$

$$I_{yz} = \int yz\,dm \qquad (5\text{-}35)$$

$$I_{zx} = \int zx\,dm$$

Parallel-Axis Theorem:

$$I_{xy} = I_{x'y'} + x_c y_c m$$

$$I_{yz} = I_{y'z'} + y_c z_c m \qquad (5\text{-}36)$$

$$I_{zx} = I_{z'x'} + z_c x_c m$$

$m$ = Total Mass of the Body.

Moment of Inertia with Respect to an Arbitrary Axis:

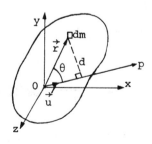

**FIGURE 5.14**

$$I_{OP} = \int d^2 \, dm = \int (\overline{u} \times \overline{r})^2 \, dm \qquad (5\text{-}37)$$

In terms of scalar quantities:

$$I_{OP} = I_x u_x^2 + I_y u_y^2 + I_z u_z^2 - 2I_{xy} u_x u_y$$
$$- 2I_{yz} u_y u_z - 2I_{zx} u_z u_x \qquad (5\text{-}38)$$

# 5.6 PRINCIPAL AXES AND INERTIAS

**AREAS**

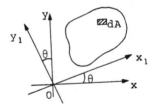

**FIGURE 5.15**

To calculate $I_{x_1}$, $I_{y_1}$, $I_{x_1 y_1}$ when $\theta$, $I_x$, $I_y$, and $I_{xy}$ are known; with respect to the inclined $x_1$ and $y_1$ axes:

$$I_{x_1} = \frac{I_x + I_y}{2} + \frac{I_x - I_y}{2} \cos 2\theta - I_{xy} \sin 2\theta$$

$$I_{y_1} = \frac{I_x + I_y}{2} - \frac{I_x - I_y}{2} \cos 2\theta - I_{xy} \sin 2\theta \qquad (5\text{-}39)$$

$$I_{x_1 y_1} = \frac{I_x - I_y}{2} \sin 2\theta + I_{xy} \cos 2\theta$$

The polar moment of inertia about the $z$ axis is independent of the orientation of $x_1$ and $y_1$:

$$J_0 = I_{x_1} + I_{y_1} = I_x + I_y \qquad (5\text{-}40)$$

Principal Axes – Axes about which the moments of inertial are maximum or minimum. The corresponding moments of inertia are called principal moments of inertia.

NOTE: Every point on the body has its own set of principal axes. The important point to consider is the centroid.

Orientation of the principal axes at the centroid; $\theta_p$ :

$$\tan 2\theta_p = \frac{-I_{xy}}{\dfrac{I_x - I_y}{2}} \qquad (5\text{-}41)$$

Equation (5-41) has two solutions, $\theta_{p_1}$ and $\theta_{p_2}$. They correspond to the maximum and minimum values of the principal moments of inertia and are 90° apart.

**PRINCIPAL MOMENTS OF INERTIA:**

$$I_{\substack{max \\ min}} = \frac{I_x + I_y}{2} \pm \sqrt{\left(\frac{I_x - I_y}{2}\right)^2 + I_{xy}^2} \qquad (5\text{-}42)$$

**NOTE**: The product of inertia with respect to the principal axes is zero. Therefore, axes of symmetry are also principal axes for a given area.

Ellipsoid of Inertia:

Equation of an ellipsoid:

$$I_x x^2 + I_y y^2 + I_z z^2 - 2I_{xy}xy - 2I_{yz}yz - 2I_{zx}zx = 1 \quad \text{(5-43)}$$

The ellipsoid defines the moment of inertia of the body with respect to any axis passing through the origin $O$. It is also called the ellipsoid of inertia at point $O$.

If the principal axes, $x_1$, $y_1$, $z_1$ are used as coordinate axes, then the equation of the ellipsoid becomes:

$$I_{x_1} x_1^2 + I_{y_1} y_1^2 + I_{z_1} z_1^2 = 1 \quad \text{(5-44)}$$

Using the principal axes, the moment of inertia with respect to an arbitrary axis through the origin $O$, equation (5-38), is:

$$I_{OP} = I_{x_1} u_{x_1}^2 + I_{y_1} u_{y_1}^2 + I_{z_1} u_{z_1}^2 \quad \text{(5-45)}$$

# 5.7 PROPERTIES OF GEOMETRIC SECTIONS

**SQUARE**
Axis of moments through center

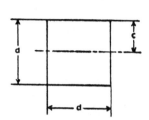

$$A = d^2$$

$$c = \frac{d}{2}$$

$$I = \frac{d^4}{12}$$

$$S = \frac{d^3}{6}$$

$$r = \frac{d}{\sqrt{12}}$$

**SQUARE**
Axis of moments on base

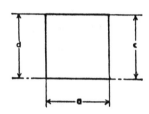

$$A = d^2$$

$$c = d$$

$$I = \frac{d^4}{3}$$

$$S = \frac{d^3}{3}$$

$$r = \frac{d}{\sqrt{3}}$$

**SQUARE**
Axis of moments on diagonal

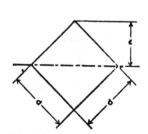

$$A = d^2$$

$$c = \frac{d}{\sqrt{2}}$$

$$I = \frac{d^4}{12}$$

$$S = \frac{d^3}{6\sqrt{2}}$$

$$r = \frac{d}{\sqrt{12}}$$

### RECTANGLE
#### Axis of moments through center

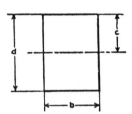

$$A = bd$$

$$c = \frac{d}{2}$$

$$I = \frac{bd^3}{12}$$

$$S = \frac{bd^2}{6}$$

$$r = \frac{d}{\sqrt{12}}$$

---

### RECTANGLE
#### Axis of moments on base

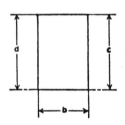

$$A = bd$$

$$c = d$$

$$I = \frac{bd^3}{3}$$

$$S = \frac{bd^2}{3}$$

$$r = \frac{d}{\sqrt{3}}$$

---

### RECTANGLE
#### Axis of moments on diagonal

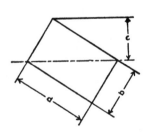

$$A = bd$$

$$c = \frac{bd}{\sqrt{b^2 + d^2}}$$

$$I = \frac{b^3 d^3}{6(b^2 + d^2)}$$

$$S = \frac{b^2 d^2}{6\sqrt{b^2 + d^2}}$$

$$r = \frac{bd}{\sqrt{6(b^2 + d^2)}}$$

### RECTANGLE
Axis of moments on any line
through center of gravity

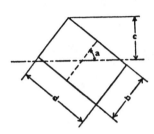

$$A = bd$$

$$c = \frac{b\sin a + d\cos a}{2}$$

$$I = \frac{bd(b^2\sin^2 a + d^2\cos^2 a)}{12}$$

$$S = \frac{bd(b^2\sin^2 a + d^2\cos^2 a)}{6(b\sin a + d\cos a)}$$

$$r = \sqrt{\frac{b^2\sin^2 a + d^2\cos^2 a}{12}}$$

### HOLLOW RECTANGLE
Axis of moments through center

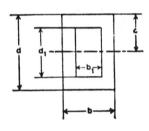

$$A = bd - b_1 d_1$$

$$c = \frac{d}{2}$$

$$I = \frac{bd^3 - b_1 d_1^3}{12}$$

$$S = \frac{bd^3 - b_1 d_1^3}{6d}$$

$$r = \sqrt{\frac{bd^3 - b_1 d_1^3}{12A}}$$

### EQUAL TRIANGLES
Axis of moments through
center of gravity

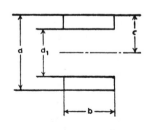

$$A = b(d - d_1)$$

$$c = \frac{d}{2}$$

$$I = \frac{b(d^3 - d_1^3)}{12}$$

$$S = \frac{b(d^3 - d_1^3)}{6d}$$

$$r = \sqrt{\frac{d^3 - d_1^3}{12(d - d_1)}}$$

## UNEQUAL RECTANGLES
### Axis of moments through center of gravity

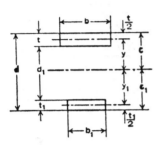

$$A = bt + b_1 t_1$$

$$c = \frac{\frac{1}{2}bt^2 + b_1 t_1 (d - \frac{1}{2}t_1)}{A}$$

$$I = \frac{bt^3}{12} + bty^2 + \frac{b_1 t_1^3}{12} + b_1 t_1 y_1^2$$

$$S = \frac{I}{c} \quad S_1 = \frac{I}{c_1}$$

$$r = \sqrt{\frac{I}{A}}$$

---

## TRIANGLE
### Axis of moments through center of gravity

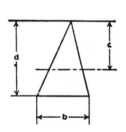

$$A = \frac{bd}{2}$$

$$c = \frac{2d}{3}$$

$$I = \frac{bd^3}{36}$$

$$S = \frac{bd^2}{24}$$

$$r = \frac{d}{\sqrt{18}}$$

---

## TRIANGLE
### Axis of moments on base

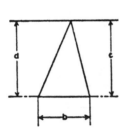

$$A = \frac{bd}{2}$$

$$c = d$$

$$I = \frac{bd^3}{12}$$

$$S = \frac{bd^2}{12}$$

$$r = \frac{d}{\sqrt{6}}$$

---

58

### TRAPEZOID
Axis of moments through
center of gravity

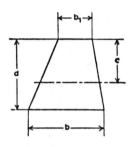

$$A = \frac{d(b + b_1)}{2}$$

$$c = \frac{d(2b + b_1)}{3(b + b_1)}$$

$$I = \frac{d^3(b^2 + 4bb_1 + b_1^2)}{36(b + b_1)}$$

$$S = \frac{d^2(b^2 + 4bb_1 + b_1^2)}{12(2b + b_1)}$$

$$r = \frac{d}{6(b + b_1)} \sqrt{2(b^2 + 4bb_1 + b_1^2)}$$

---

### CIRCLE
Axis of moments
through center

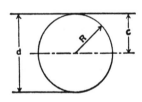

$$A = \frac{\pi d^2}{4} = \pi R^2$$

$$c = \frac{d}{2} \quad = R$$

$$I = \frac{\pi d^4}{64} = \frac{\pi R^2}{4}$$

$$S = \frac{\pi d^3}{32} = \frac{\pi R^3}{4}$$

$$r = \frac{d}{4} \quad = \frac{R}{2}$$

---

### HOLLOW CIRCLE
Axis of moments
through center

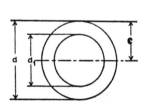

$$A = \frac{\pi(d^2 - d_1^2)}{4}$$

$$c = \frac{d}{2}$$

$$I = \frac{\pi(d^4 - d_1^4)}{64}$$

$$S = \frac{\pi(d^4 - d_1^4)}{32d}$$

$$r = \frac{\sqrt{d^2 + d_1^2}}{4}$$

---

### HALF CIRCLE
Axis of moments through
center of gravity

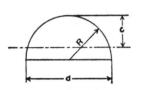

$$A = \frac{\pi R^2}{2}$$

$$c = R\left(1 - \frac{4}{3\pi}\right)$$

$$I = R^4\left(\frac{\pi}{8} - \frac{8}{9\pi}\right)$$

$$S = \frac{R^3}{24}\frac{(9\pi^2 - 64)}{(3\pi - 4)}$$

$$r = R\frac{\sqrt{9\pi^2 - 64}}{6\pi}$$

---

### PARABOLA

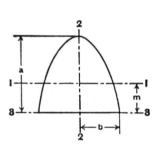

$$A = \frac{4}{3}ab$$

$$m = \frac{2}{5}a$$

$$I_1 = \frac{16}{175}a^3b$$

$$I_2 = \frac{4}{15}ab^3$$

$$I_3 = \frac{32}{105}a^3b$$

---

### HALF PARABOLA

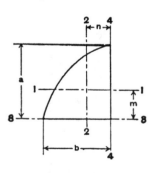

$$A = \frac{2}{3}ab$$

$$m = \frac{2}{5}a$$

$$n = \frac{3}{8}b$$

$$I_1 = \frac{8}{175}a^3b, \quad I_2 = \frac{19}{480}ab^3$$

$$I_3 = \frac{16}{105}a^3b, \quad I_4 = \frac{2}{15}ab^3$$

## COMPLEMENT OF HALF PARABOLA

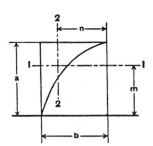

$$A = \frac{1}{3}\, ab$$

$$m = \frac{7}{10}\, a$$

$$n = \frac{3}{4}\, b$$

$$I_1 = \frac{37}{2100}\, a^3 b$$

$$I_2 = \frac{1}{80}\, ab^3$$

## PARABOLIC FILLET IN RIGHT ANGLE

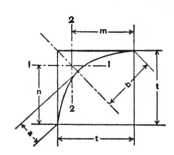

$$a = \frac{t}{2\sqrt{2}}$$

$$b = \frac{t}{\sqrt{2}}$$

$$A = \frac{1}{6}\, t^2$$

$$m = n = \frac{4}{5}\, t$$

$$I_1 = I_2 = \frac{11}{2100}\, t^4$$

## * HALF ELLIPSE

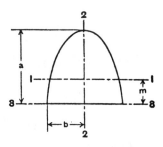

$$A = \frac{1}{2}\, \pi ab$$

$$m = \frac{4a}{3\pi}$$

$$I_1 = a^3 b \left( \frac{\pi}{8} - \frac{8}{9\pi} \right)$$

$$I_2 = \frac{1}{8}\, \pi ab^3$$

$$I_3 = \frac{1}{8}\, \pi a^3 b$$

* To obtain properties of half circle, quarter circle and circular complement substitute $a = b = R$.

61

## *QUARTER ELLIPSE

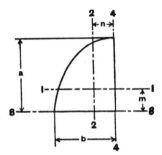

$$A = \frac{1}{4}\pi ab$$

$$m = \frac{4a}{3\pi}$$

$$n = \frac{4b}{3\pi}$$

$$I_1 = a^3b\left(\frac{\pi}{16} - \frac{4}{9\pi}\right), \quad I_3 = \frac{1}{16}\pi a^3 b$$

$$I_2 = ab^3\left(\frac{\pi}{16} - \frac{4}{9\pi}\right), \quad I_4 = \frac{1}{16}\pi ab^3$$

## *ELLIPTIC COMPLEMENT

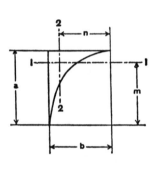

$$A = ab\left(1 - \frac{\pi}{4}\right)$$

$$m = \frac{a}{6\left(1 - \frac{\pi}{4}\right)}$$

$$n = \frac{b}{6\left(1 - \frac{\pi}{4}\right)}$$

$$I_1 = a^3b\left(\frac{1}{3} - \frac{\pi}{16} - \frac{1}{36\left(1 - \frac{\pi}{4}\right)}\right)$$

$$I_2 = ab^3\left(\frac{1}{3} - \frac{\pi}{16} - \frac{1}{36\left(1 - \frac{\pi}{4}\right)}\right)$$

## REGULAR POLYGON
### Axis of moments through center

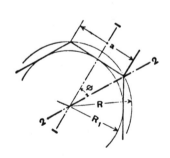

$n = $ Number of sides, $\qquad \phi = \dfrac{180°}{n}$

$$a = 2\sqrt{R^2 - R_1^2}$$

$$R = \frac{a}{2\sin\phi}, \qquad R_1 = \frac{a}{2\tan\phi}$$

$$A = \frac{1}{4}na^2\cot\phi = \frac{1}{2}nR^2\sin 2\phi = nR_1^2\tan\phi$$

$$I_1 = I_2 = \frac{A(6R^2 - a^2)}{24} = \frac{A(12R_1^2 + a^2)}{48}$$

$$r_1 = r_2 = \sqrt{\frac{6R^2 - a^2}{24}} = \sqrt{\frac{12R_1^2 + a^2}{48}}$$

* To obtain properties of half circle, quarter circle and circular complement substitute $a = b = R$.

62

## ANGLE
Axis of moments through
center of gravity

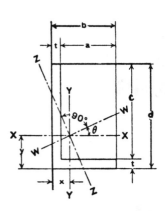

Z–Z is axis of minimum *I*

$$\tan 2\phi = \frac{2K}{I_y - I_x}, \ A = t(b+c)$$

$$x = \frac{b^2 + ct}{2(b+c)}, \quad y = \frac{d^2 + at}{2(b+c)}$$

$K =$ Product of Inertia about
$\quad\quad X - X \ \& \ Y - Y$

$$= \pm \frac{abcdt}{4(b+c)}$$

$$I_x = \tfrac{1}{3}\left( t(d-y)^3 + by^3 - a(y-t)^3 \right)$$

$$I_y = \tfrac{1}{3}\left( t(b-x)^3 + dx^3 - c(x-t)^3 \right)$$

$$I_z = I_x \sin^2\theta + I_y \cos^2\theta + K \sin 2\theta$$

$$I_w = I_x \cos^2\theta + I_y \sin^2\theta - K \sin 2\theta$$

*K* is negative when heel of angle,
with respect to *c. g.*, is in 1st or 3rd
quadrant, positive when in 2nd or
4th quadrant.

## BEAMS AND CHANNELS
Transverse force oblique
through center of gravity

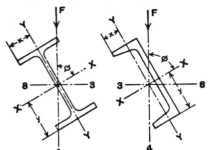

$$I_3 = I_x \sin^2\phi + I_Y \cos^2\phi$$

$$I_4 = I_x \cos^2\phi + I_Y \sin^2\phi$$

$$f = M\left( \frac{Y}{I_x} \sin\phi + \frac{x}{I_Y} \cos\phi \right)$$

where *M* is bending moment due
to force *F*.

# CHAPTER 6

# STRESSES IN BEAMS

## 6.1 BENDING STRESS DISTRIBUTION

A beam deforms as a result of the action of a bending moment. During the deformed condition, the fibers on the outer surface (convex) of the beam are in tension and the fibers on the inner surface (concave) of the beam are in compression. The stress distribution across the cross-section of a deformed beam under bending is shown in Figure 6.1. The locus of all those points inside the beam at which the stress is zero, is called the neutral axis.

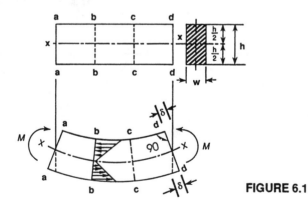

**FIGURE 6.1**

By the stress distribution diagram, it is clear that the bending stresses increase linearly from zero at the neutral plane to a maximum at the outer fibers. The compressive stress in the top fibers are equal to the tensile stress in the bottom fibers following Hooke's law, that the stress is proportional to deformation.

## 6.2   FLEXURE FORMULA

The following assumptions are based on the relationship between bending moment $M$ and the resulting bending stress $\sigma_b$, also known as flexure formula.

1.  Transverse planes remain transverse before and after the bending, i.e. no warping occurs.

2.  The beam has a homogenous and isotropic material which obeys Hooke's law. $E$ is the same for tension and compression.

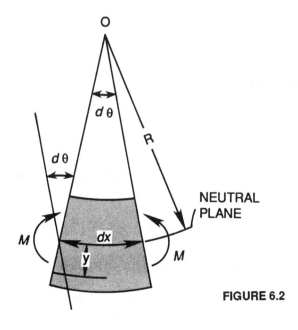

FIGURE 6.2

65

3. The beam is subjected to a pure bending moment i.e., there is no twisting or buckling load on the beam.

4. The beam is straight with a rectangular cross-section.

From Figure 6.2, the strain in a fiber located at a depth $Y$ from the neutral axis is,

$$\varepsilon_x = \frac{dl}{dx} \tag{6-1}$$

where the bending deformation $dl = yd\theta$

$$\therefore \quad \varepsilon_x = \frac{yd\theta}{dx} \tag{6-2}$$

Using similar triangles approach

$$\frac{yd\theta}{dx} = \frac{y}{R} \Rightarrow \frac{d\theta}{dx} = \frac{1}{R} \tag{6-3}$$

$$\tag{6-4}$$
$$\therefore \quad \varepsilon_x = \frac{y}{R}$$

Using Hooke's law,

$$\sigma_x = \varepsilon_x E = \frac{E}{R} y \tag{6-5}$$

Now, if $dF$ is the small increment of force acting on an elemental area $dA$, then

$$dF = \sigma_x dA \tag{6-6}$$

The corresponding infinitesimal moment is

$$dM = dF \cdot y = (\sigma_x dA) \cdot y$$

or

$$dM = \left(\frac{E}{R}\right) y^2 \, dA \qquad (6\text{-}7)$$

on integration, the total resisting moment is,

$$M = \int \left(\frac{E}{R}\right) y^2 \, dA = \frac{E}{R} \int y^2 \, dA \qquad (6\text{-}8)$$

where $\int y^2 \, dA = I$ is called the second moment of area or moment of inertia of the area of cross-section with respect to the neutral axis.

Thus

$$M = \frac{E}{R} I \quad \text{or} \quad \frac{M}{I} = \frac{E}{R} \qquad (6\text{-}9)$$

since,

$$\sigma_x = \left(\frac{E}{R}\right) y \implies \frac{\sigma_x}{y} = \frac{E}{R}$$

Therefore,

$$\boxed{\frac{M}{I} = \frac{\sigma_x}{y} = \frac{E}{R}} \qquad (6\text{-}10)$$

This equation is known as flexure formula and is extensively used in the design of structural members.

When $c$ is the distance of the extreme fiber from the neutral axis, $\sigma_x = \sigma_{max}$

$$\boxed{\therefore \quad M = \sigma_{max} \frac{I}{c}} \qquad (6\text{-}11)$$

where $I/c$ is called section modulus of the beam.

# 6.3  SHEAR STRESSES IN BEAMS

In a beam of rectangular cross-section, both the horizontal and the vertical shearing stresses vary parabolically. The general equation for shear stress in a beam is,

$$\tau = \frac{VQ}{Ib}$$

(6-12)

**Shear Formula**

where  $\tau$   = Shear stress
$V$   = Total transverse shear
$I$    = Second moment of area
$Q$   = First moment of area

Now

$$Q = \int_A y \, dA$$

(6-13)

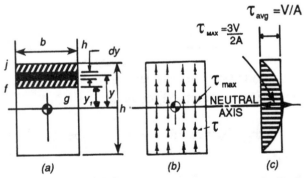

FIGURE 6.3—(a) beam cross-sectional area, (b) longitudinal cut, (c) shear stress distribution.

From Figure 6.3 $dA = bdy$

68

$$\therefore Q = \int y\, dA = \int by\, dy = b\int_{y_1}^{h/2} y\, dy = \frac{by^2}{2}\bigg|_{y_1}^{h/2}$$

$$Q = \frac{b}{2}\left[\left(\frac{h}{2}\right)^2 - y_1^2\right] \qquad (6\text{-}14)$$

Therefore, in a beam of rectangular cross-section, the shear stress varies parabolically. The maximum value of a shearing stress $\tau = \tau_{max}$ is obtained when $y_1 = 0$ or at the neutral axis. At the extreme fibers, $y_1 = \pm h/2$, therefore $Q = 0$ thus making $\tau = 0$. At increasing distances from the neutral axis, shear stress gradually diminishes as shown in Figure 6.3(b).

The direction of the shearing stresses at the section through a beam is the same as that of the shearing force $V$. This fact may be used to determine the sense of the shearing stresses.

As noted above, the maximum shearing stress in a rectangular beam occurs at the neutral axis, and for this case the general expression for $\tau_{max}$ may be simplified since $y_1 = 0$.

$$\frac{Vh^2}{8I} = \frac{Vh^2}{\dfrac{8bh^3}{12}} = \frac{3}{2}\frac{V}{bh} = \frac{3}{2}\frac{V}{A} \qquad (6\text{-}15)$$

## 6.4    SHEAR CENTER IN BEAMS

If a structural member does not have a longitudinal plane of symmetry or if the loads are not acting in the plane of symmetry, the member will twist. Bending without twisting (pure bending) can be made to occur if the loads are applied in the same plane in which the resultant of the shear stresses acts. In other words, a pure bending is achieved only when the resultant shear force passes through the shear center. Therefore, a shear center is defined as a point in the cross-section of a member through which the resultant of the transverse shearing stresses must act, regardless of the plane of transverse loads so that pure bending could result.

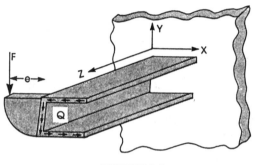

**FIGURE 6.4**

A shear force in any portion of a cross section can be found from $dF = \tau\, dA$. The internal shear forces, $Q$, in the flanges of the channel of Figure 6.4 tend to produce clockwise torsion in the channel cross-section. To avoid torsion, the applied load $F$ should intersect through a point distance $e$ from the channel. Generally speaking, this point is known as the **shear center.** The distance $e$ is called the eccentricity of the load.

The following illustrations will explain how the shear center is determined.

Given a cantilever beam with a channel cross-section (Figure 6.5(a)) and a force applied as shown, it is required to compute $e$.

In Figure 6.5(b) we have shown a free body exposing a section at $x$ having a normal in the plus $x$ coordinate direction. Finally, in Figure 6.5(c) this section is shown with a shear-stress distribution corresponding to a positive shear force $V_y$. For the

70

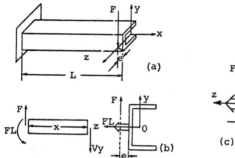
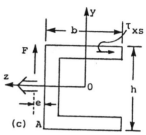

**FIGURE 6.5**

free body shown in Figure 6.5(b) the applied shear loading system is the force $F$ and the couple $FL$. To determine $e$, we set the twisting moment of the applied shear-force system about an axis parallel to the $x$ axis and going through point $A$ (see Figure 6.5(c)) equal and opposite to the moment about this axis of the shear-stress distribution of the section at position $x$. For such an axis only the stresses in the upper flange contribute moment, because the other stresses have no moment arm with respect to $A$. Accordingly, we can say, using the proper directional signs, for the moments that

$$[-(Fe)] = -\left[\int_0^b \{[-(h-t_1)\tau_{xs}\}t_1\,ds\right] \quad (6\text{-}16)$$

Note that the couple $FL$ has no moment about the axis at $A$ since it is orthogonal to this axis. Employing

$$(t_{xs})_{\text{flange}} = \frac{V_y s \dfrac{h-t_1}{2}}{I_{zz}}$$

for $\tau_{xs}$ we obtain

$$-[Fe] = \int_0^b t_1 \frac{V_y(h-t_1)^2}{2I_{zz}}s\,ds$$

71

$$= \frac{t_1 V_y (h - t_1)^2}{2 I_{zz}} s_2^2 \Big|_0^b = \frac{t_1 V_y (h - t_1)^2 b^2}{4 I_{zz}} \qquad (6\text{-}17)$$

Since $V_y = -F$, for $e$ we obtain

$$e = \frac{t_1 (h - t_1)^2 b^2}{4 I_{zz}} \qquad (6\text{-}18)$$

Since $t_1$ and $t_2$ are small compared to $b$, we shall approximate $I_{zz}$ employing

$$I_{zz} = b t_1 \left( h - t_1 \right) \left( \frac{h}{2} - \frac{t_1}{2} \right) + \frac{1}{12} t_2 \left( h - t_1 \right)^3 + \frac{1}{6} b t_1^3$$

for this purpose. We may thus say that for $t \ll h$ that

$$I_{zz} \approx \frac{b t_1 h^2}{2} + \frac{1}{12} t_2 h^3 \qquad (6\text{-}19)$$

$$\therefore \quad e = \frac{t_1 (h - t_1)^2 b^2}{4} \cdot \frac{12}{6 b t_1 h^2 + t_2 h^3}$$

dropping $t_1$ in the squared bracket

$$e = \frac{t_1 h^2 b^2}{1} \times \frac{3}{6 b t_1 h^2 + t_2 h^3} = \frac{t_1 b^2}{2 b t_1 + \frac{t_2 h}{3}} \qquad (6\text{-}20)$$

$$\therefore \quad e = \frac{t_1 b^2}{2 b t_1 + \frac{t_2 h}{3}}$$

# CHAPTER 7

# DESIGN OF BEAMS

## 7.1 BEAM DESIGN CRITERIA

**DESIGN FACTORS**

- Loading (arrangement and magnitude)
- Span of beams
- Type of support
- Permissible stress
- Permissible deflection
- Size and shape of beam

**DESIGN STEPS**

- Analyze the loads (dynamic and static) on beams, unsupported span and type of support, in detail.

- Estimate bending moment, shearing force, twisting moment, etc.

- Choose the factor of safety, depending on the accuracy of assumptions made for the above estimations.

73

— Select the material of beam having the allowable stress and deflection within permissible limits.

— Select the size and shape of the beam which will given an optimum section modulus.

In addition to the above factors, we have to keep fatigue, creep and vibrational criteria for design in mind. Moreover, the design should be economical.

Rupture strength of the material in bending is found by Equation (7-1)

$$\sigma_r = \frac{M_r \overline{y}}{I}$$

(7-1)

where  $\sigma_r$  = modulus of rupture
$M_r$  = moment at the breaking point
$I$  = moment of inertia of beam sections.

## 7.1.1    SHEAR FORCE

**VERTICAL SHEAR FORCE**

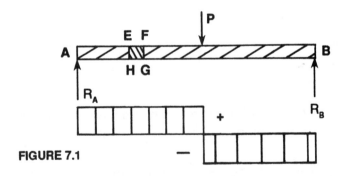

**FIGURE 7.1**

Vertical shear force at a section is the algebraic sum of all vertical forces acting on the beam to one end of that section.

74

## HORIZONTAL SHEAR FORCE

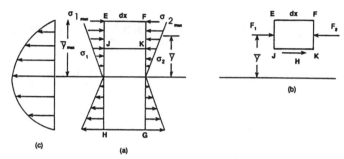

**FIGURE 7.2**—(a) Enlarged section *EFGH*, (b) horizontal force on section *EFGH*, (c) horizontal shear stress distribution.

From Figure 7.2(b), Horizontal Shear force:

$$H = F_2 - F_1$$

Shear stress

$$\tau_H = \frac{Va\overline{y}}{Ib} \qquad (7\text{-}2)$$

where  $\tau_H$  = Horizontal shear stress
  $V$  = Vertical shear force at the section
  $a$  = area of cross section of beam between fiber being considered and nearest extreme fiber
  $\overline{y}$  = distance from neutral axis of entire cross section to centroid of area, $a$.
  $I$  = moment of inertia of entire cross section of beam
  $b$  = width of cross section at fiber being considered

**NOTE**: At any point in a member subjected to shearing forces, there exists equal shearing stresses in planes mutually perpendicular.

75

Beam diagrams and formulas for various static and moving loads are given in Section 7.5.

# 7.2     BEAM DESIGN PROCEDURES

The most general procedures of beam design are given as follows:

**DESIGN PROCEDURE 1:**

By trying different values of $b$ to solve the section modulus equation, $z = \dfrac{bh^2}{6}$ , a table of results could be formed. Through this procedure, only a few of the suitable beams, which can satisfy the requirements of the strength, are picked up out of several beam sizes available. Further rational constraints cause the final selection of the most suitable beam size.

**DESIGN PROCEDURE 2:**

Using a pre-determined width-depth ratio $(b/t)$, generally $^1/_3$, $^1/_2$, $^2/_3$ or $^3/_4$, and selecting the most appropriate ratio, one can solve the equation by having two unknown variables, $b$ and $h$. The selection of $b/t$ ratio is totally up to the designer's discretion. Any such pre-calculated ratio can vary in size, in either variable $b$ or $t$, in the beam finally selected.

**DESIGN PROCEDURE 3:**

The values of the section modulus and the area are determined first. Thereafter, a suitable selection of a beam is made from the tables of beam sizes and beam properties such as moment of inertia, section modulus, area and weight.

# 7.3   PLASTIC ANALYSIS OF BEAMS

Materials that deform without fracture on further applications of load beyond the yield point are called plastic materials. Plasticity, therefore, is the ability of a material to deform non-elastically without having a rupture. (Refer to Section 9.6 for more details.)

# 7.4   ECONOMY IN BEAM DESIGN

A beam with a minimum cross-sectional area and satisfying all strength requirements within the specified constraints, is the most economic. Generally, for the same strength criteria, a beam of greater depth is more economical than one with lesser depth. Higher beam depth increases the moment of inertia and hence, the section modulus which in turn reduces the stress developed in the material at the same loading conditions.

The arrangement of supports, selection of material, and proper analysis of design criteria make a design economically sound.

# 7.5 BEAM DIAGRAMS AND FORMULAS

## 7.5.1 FOR VARIOUS STATIC LOADING CONDITIONS

### 1. SIMPLE BEAM—UNIFORMLY DISTRIBUTED LOAD

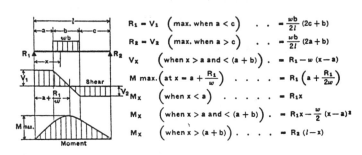

$$R_1 = V_1 \quad \left( \text{max. when } a < c \right) \quad \ldots \quad = \frac{wb}{2l}(2c+b)$$

$$R_2 = V_2 \quad \left( \text{max. when } a > c \right) \quad \ldots \quad = \frac{wb}{2l}(2a+b)$$

$$V_x \quad \left( \text{when } x > a \text{ and } < (a+b) \right) . \quad = R_1 - w(x-a)$$

$$M \text{ max.} \left( \text{at } x = a + \frac{R_1}{w} \right) \ldots \ldots = R_1\left(a + \frac{R_1}{2w}\right)$$

$$M_x \quad \left( \text{when } x < a \right) \ldots \ldots = R_1 x$$

$$M_x \quad \left( \text{when } x > a \text{ and } < (a+b) \right) . \quad = R_1 x - \frac{w}{2}(x-a)^2$$

$$M_x \quad \left( \text{when } x > (a+b) \right) \ldots \ldots = R_2(l-x)$$

### 2. SIMPLE BEAM—LOAD INCREASING UNIFORMLY TO ONE END

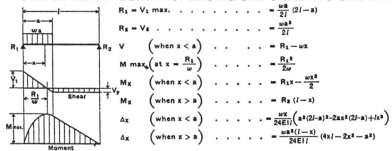

$$R_1 = V_1 \text{ max.} \ldots \ldots \ldots = \frac{wa}{2l}(2l-a)$$

$$R_2 = V_2 \ldots \ldots \ldots = \frac{wa^2}{2l}$$

$$V \quad \left( \text{when } x < a \right) \ldots \ldots = R_1 - wx$$

$$M \text{ max.} \left( \text{at } x = \frac{R_1}{w} \right) \ldots \ldots = \frac{R_1^2}{2w}$$

$$M_x \quad \left( \text{when } x < a \right) \ldots \ldots = R_1 x - \frac{wx^2}{2}$$

$$M_x \quad \left( \text{when } x > a \right) \ldots \ldots = R_2(l-x)$$

$$\Delta_x \quad \left( \text{when } x < a \right) \ldots \ldots = \frac{wx}{24EIl}\left(a^2(2l-a)^2 - 2ax^2(2l-a) + lx^3\right)$$

$$\Delta_x \quad \left( \text{when } x > a \right) \ldots \ldots = \frac{wa^2(l-x)}{24EIl}(4xl - 2x^2 - a^2)$$

### 3. SIMPLE BEAM—LOAD INCREASING UNIFORMLY TO CENTER

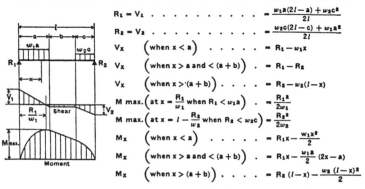

$$R_1 = V_1 \ldots \ldots \ldots = \frac{w_1 a(2l-a) + w_2 c^2}{2l}$$

$$R_2 = V_2 \ldots \ldots \ldots = \frac{w_2 c(2l-c) + w_1 a^2}{2l}$$

$$V_x \quad \left( \text{when } x < a \right) \ldots \ldots = R_1 - w_1 x$$

$$V_x \quad \left( \text{when } x > a \text{ and } < (a+b) \right) . \quad = R_1 - R_2$$

$$V_x \quad \left( \text{when } x > (a+b) \right) \ldots \ldots = R_2 - w_2(l-x)$$

$$M \text{ max.} \left( \text{at } x = \frac{R_1}{w_1} \text{ when } R_1 < w_1 a \right) . \quad = \frac{R_1^2}{2w_1}$$

$$M \text{ max.} \left( \text{at } x = l - \frac{R_2}{w_2} \text{ when } R_2 < w_2 c \right) = \frac{R_2^2}{2w_2}$$

$$M_x \quad \left( \text{when } x < a \right) \ldots \ldots = R_1 x - \frac{w_1 x^2}{2}$$

$$M_x \quad \left( \text{when } x > a \text{ and } < (a+b) \right) . \quad = R_1 x - \frac{w_1 a}{2}(2x-a)$$

$$M_x \quad \left( \text{when } x > (a+b) \right) \ldots \ldots = R_2(l-x) - \frac{w_2(l-x)^2}{2}$$

# SIMPLE BEAM—UNIFORM LOAD PARTIALLY DISTRIBUTED

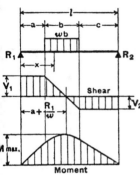

$$R_1 = V_1 \quad \left(\text{max. when } a < c\right) \quad \cdots \quad = \frac{wb}{2l}(2c + b)$$

$$R_2 = V_2 \quad \left(\text{max. when } a > c\right) \quad \cdots \quad = \frac{wb}{2l}(2a + b)$$

$$V_x \quad \left(\text{when } x > a \text{ and } < (a + b)\right) . \quad = R_1 - w(x - a)$$

$$M \text{ max.}\left(\text{at } x = a + \frac{R_1}{w}\right) \quad \cdots \quad = R_1\left(a + \frac{R_1}{2w}\right)$$

$$M_x \quad \left(\text{when } x < a\right) \quad \cdots \quad = R_1 x$$

$$M_x \quad \left(\text{when } x > a \text{ and } < (a + b)\right) . \quad = R_1 x - \frac{w}{2}(x - a)^2$$

$$M_x \quad \left(\text{when } x > (a + b)\right) \quad \cdots \quad = R_2(l - x)$$

# SIMPLE BEAM—UNIFORM LOAD PARTIALLY DISTRIBUTED AT ONE END

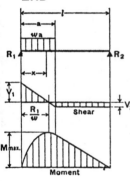

$$R_1 = V_1 \text{ max.} \quad \cdots \quad = \frac{wa}{2l}(2l - a)$$

$$R_2 = V_2 \quad \cdots \quad = \frac{wa^2}{2l}$$

$$V \quad \left(\text{when } x < a\right) \quad \cdots \quad = R_1 - wx$$

$$M \text{ max.}\left(\text{at } x = \frac{R_1}{w}\right) \quad \cdots \quad = \frac{R_1^2}{2w}$$

$$M_x \quad \left(\text{when } x < a\right) \quad \cdots \quad = R_1 x - \frac{wx^2}{2}$$

$$M_x \quad \left(\text{when } x > a\right) \quad \cdots \quad = R_2(l - x)$$

$$\Delta_x \quad \left(\text{when } x < a\right) \quad \cdots \quad = \frac{wx}{24EIl}\left(a^2(2l - a)^2 - 2ax^2(2l - a) + lx^3\right)$$

$$\Delta_x \quad \left(\text{when } x > a\right) \quad \cdots \quad = \frac{wa^2(l - x)}{24EIl}(4xl - 2x^2 - a^2)$$

# SIMPLE BEAM—UNIFORM LOAD PARTIALLY DISTRIBUTED AT EACH END

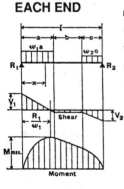

$$R_1 = V_1 \quad \cdots \quad = \frac{w_1 a(2l - a) + w_2 c^2}{2l}$$

$$R_2 = V_2 \quad \cdots \quad = \frac{w_2 c(2l - c) + w_1 a^2}{2l}$$

$$V_x \quad \left(\text{when } x < a\right) \quad \cdots \quad = R_1 - w_1 x$$

$$V_x \quad \left(\text{when } x > a \text{ and } < (a + b)\right) . \quad = R_1 - R_2$$

$$V_x \quad \left(\text{when } x > (a + b)\right) \quad \cdots \quad = R_2 - w_2(l - x)$$

$$M \text{ max.}\left(\text{at } x = \frac{R_1}{w_1} \text{ when } R_1 < w_1 a\right) \quad = \frac{R_1^2}{2w_1}$$

$$M \text{ max.}\left(\text{at } x = l - \frac{R_2}{w_2} \text{ when } R_2 < w_2 c\right) = \frac{R_2^2}{2w_2}$$

$$M_x \quad \left(\text{when } x < a\right) \quad \cdots \quad = R_1 x - \frac{w_1 x^2}{2}$$

$$M_x \quad \left(\text{when } x > a \text{ and } < (a + b)\right) . \quad = R_1 x - \frac{w_1 a}{2}(2x - a)$$

$$M_x \quad \left(\text{when } x > (a + b)\right) \quad \cdots \quad = R_2(l - x) - \frac{w_2(l - x)^2}{2}$$

## 7. SIMPLE BEAM—CONCENTRATED LOAD AT CENTER

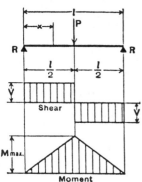

Equivalent Tabular Load . . . . . $= 2P$

$R = V$ . . . . . . . . $= \dfrac{P}{2}$

M max. $\left(\text{at point of load}\right)$ . . . . . $= \dfrac{Pl}{4}$

$M_x$ $\left(\text{when } x < \dfrac{l}{2}\right)$ . . . . . $= \dfrac{Px}{2}$

$\Delta$max. $\left(\text{at point of load}\right)$ . . . . . $= \dfrac{Pl^3}{48EI}$

$\Delta_x$ $\left(\text{when } x < \dfrac{l}{2}\right)$ . . . . . $= \dfrac{Px}{48EI}(3l^2 - 4x^2)$

## 8. SIMPLE BEAM—CONCENTRATED LOAD AT ANY POINT

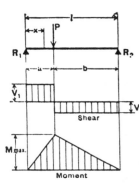

Equivalent Tabular Load . . . . . $= \dfrac{8\,Pab}{l^2}$

$R_1 = V_1\left(\text{max. when } a < b\right)$ . . . . $= \dfrac{Pb}{l}$

$R_2 = V_2\left(\text{max. when } a > b\right)$ . . . . $= \dfrac{Pa}{l}$

M max. $\left(\text{at point of load}\right)$ . . . . $= \dfrac{Pab}{l}$

$M_x$ $\left(\text{when } x < a\right)$ . . . . $= \dfrac{Pbx}{l}$

$\Delta$max. $\left(\text{at } x = \sqrt{\dfrac{a(a+2b)}{3}} \text{ when } a > b\right)$ $= \dfrac{Pab(a+2b)\sqrt{3a(a+2b)}}{27\,EI\,l}$

$\Delta_a$ $\left(\text{at point of load}\right)$ . . . . $= \dfrac{Pa^2b^2}{3EI\,l}$

$\Delta_x$ $\left(\text{when } x < a\right)$ . . . . $= \dfrac{Pbx}{6EI\,l}(l^2 - b^2 - x^2)$

## 9. SIMPLE BEAM—TWO EQUAL CONCENTRATED LOADS SYMMETRICALLY PLACED

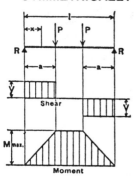

Equivalent Tabular Load . . . . . $= \dfrac{8\,Pa}{l}$

$R = V$ . . . . . . . . . . . $= P$

M max. $\left(\text{between loads}\right)$ . . . . . $= Pa$

$M_x$ $\left(\text{when } x < a\right)$ . . . . . $= Px$

$\Delta$max. $\left(\text{at center}\right)$ . . . . . . . $= \dfrac{Pa}{24EI}(3l^2 - 4a^2)$

$\Delta_x$ $\left(\text{when } x < a\right)$ . . . . . . $= \dfrac{Px}{6EI}(3la - 3a^2 - x^2)$

$\Delta_x$ $\left(\text{when } x > a \text{ and } < (l-a)\right)$ . . $= \dfrac{Pa}{6EI}(3lx - 3x^2 - a^2)$

## 30. SIMPLE BEAM—TWO EQUAL CONCENTRATED LOADS UNSYMMETRICALLY PLACED

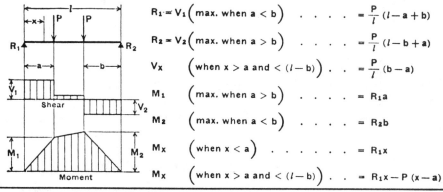

$$R_1 = V_1 \left(\text{max. when } a < b\right) \quad \ldots \quad = \frac{P}{l}\,(l - a + b)$$

$$R_2 = V_2 \left(\text{max. when } a > b\right) \quad \ldots \quad = \frac{P}{l}\,(l - b + a)$$

$$V_x \quad \left(\text{when } x > a \text{ and } < (l - b)\right) \quad \ldots \quad = \frac{P}{l}\,(b - a)$$

$$M_1 \quad \left(\text{max. when } a > b\right) \quad \ldots \quad = R_1 a$$

$$M_2 \quad \left(\text{max. when } a < b\right) \quad \ldots \quad = R_2 b$$

$$M_x \quad \left(\text{when } x < a\right) \quad \ldots \quad = R_1 x$$

$$M_x \quad \left(\text{when } x > a \text{ and } < (l - b)\right) \quad \ldots \quad = R_1 x - P\,(x - a)$$

## 31. SIMPLE BEAM—TWO UNEQUAL CONCENTRATED LOADS UNSYMMETRICALLY PLACED

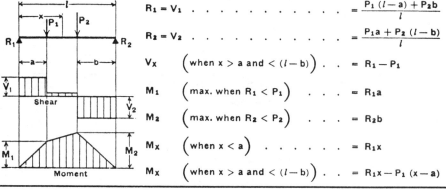

$$R_1 = V_1 \quad \ldots \quad = \frac{P_1\,(l - a) + P_2 b}{l}$$

$$R_2 = V_2 \quad \ldots \quad = \frac{P_1 a + P_2\,(l - b)}{l}$$

$$V_x \quad \left(\text{when } x > a \text{ and } < (l - b)\right) \quad \ldots \quad = R_1 - P_1$$

$$M_1 \quad \left(\text{max. when } R_1 < P_1\right) \quad \ldots \quad = R_1 a$$

$$M_2 \quad \left(\text{max. when } R_2 < P_2\right) \quad \ldots \quad = R_2 b$$

$$M_x \quad \left(\text{when } x < a\right) \quad \ldots \quad = R_1 x$$

$$M_x \quad \left(\text{when } x > a \text{ and } < (l - b)\right) \quad \ldots \quad = R_1 x - P_1\,(x - a)$$

## 32. BEAM FIXED AT ONE END, SUPPORTED AT OTHER—UNIFORMLY DISTRIBUTED LOAD

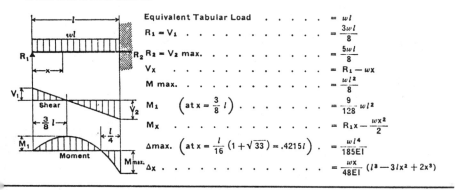

$$\text{Equivalent Tabular Load} \quad \ldots \quad = wl$$

$$R_1 = V_1 \quad \ldots \quad = \frac{3wl}{8}$$

$$R_2 = V_2 \text{ max.} \quad \ldots \quad = \frac{5wl}{8}$$

$$V_x \quad \ldots \quad = R_1 - wx$$

$$M \text{ max.} \quad \ldots \quad = \frac{wl^2}{8}$$

$$M_1 \quad \left(\text{at } x = \frac{3}{8}\,l\right) \quad \ldots \quad = \frac{9}{128}\,wl^2$$

$$M_x \quad \ldots \quad = R_1 x - \frac{wx^2}{2}$$

$$\Delta \text{max.} \quad \left(\text{at } x = \frac{l}{16}\,(1 + \sqrt{33}) = .4215l\right) \quad = \frac{wl^4}{185EI}$$

$$\Delta_x \quad \ldots \quad = \frac{wx}{48EI}\,(l^3 - 3lx^2 + 2x^3)$$

81

## 13. BEAM FIXED AT ONE END, SUPPORTED AT OTHER—CONCENTRATED LOAD AT CENTER

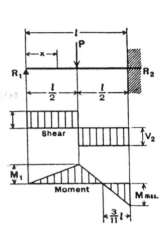

Equivalent Tabular Load $\quad . \quad . \quad = \dfrac{3P}{2}$

$R_1 = V_1 \quad . \quad . \quad . \quad . \quad . \quad . \quad = \dfrac{5P}{16}$

$R_2 = V_2 \text{ max.} \quad . \quad . \quad . \quad . \quad . \quad = \dfrac{11P}{16}$

$M \text{ max.} \left( \text{at fixed end} \right) \quad . \quad . \quad . = \dfrac{3Pl}{16}$

$M_1 \quad \left( \text{at point of load} \right) . \quad . \quad = \dfrac{5Pl}{32}$

$M_x \quad \left( \text{when } x < \dfrac{l}{2} \right) \quad . \quad . \quad . = \dfrac{5Px}{16}$

$M_x \quad \left( \text{when } x > \dfrac{l}{2} \right) \quad . \quad . \quad . = P \left( \dfrac{l}{2} - \dfrac{11x}{.16} \right)$

$\Delta \text{max.} \left( \text{at } x = l \sqrt{\dfrac{1}{5}} = .4472l \right) = \dfrac{Pl^3}{48EI\sqrt{5}} = .009317 \dfrac{Pl^3}{EI}$

$\Delta_x \quad \left( \text{at point of load} \right) . \quad . \quad = \dfrac{7Pl3}{768EI}$

$\Delta_x \quad \left( \text{when } x < \dfrac{l}{2} \right) \quad . \quad . \quad . = \dfrac{Px}{96EI} (3l^2 - 5x^2)$

$\Delta^x \quad \left( \text{when } x > \dfrac{l}{2} \right) \quad . \quad . \quad . = \dfrac{P}{96EI} (x - l)^2 (11x - 2l)$

## 14. BEAM FIXED AT ONE END, SUPPORTED AT OTHER—CONCENTRATED LOAD AT ANY POINT

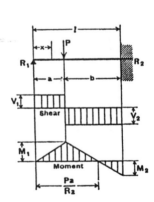

$R_1 = V_1 \quad . \quad . \quad . \quad . \quad . \quad . \quad . = \dfrac{Pb^2}{2l^3} (a + 2l)$

$R_2 = V_2 \quad . \quad . \quad . \quad . \quad . \quad . \quad . = \dfrac{Pa}{2l^3} (3l^2 - a^2)$

$M_1 \quad \left( \text{at point of load} \right) . \quad . \quad . \quad . = R_1 a$

$M_2 \quad \left( \text{at fixed end} \right) . \quad . \quad . \quad . \quad . = \dfrac{Pab}{2l^2} (a + l)$

$M_x \quad \left( \text{when } x < a \right) . \quad . \quad . \quad . \quad . = R_1 x$

$M_x \quad \left( \text{when } x > a \right) . \quad . \quad . \quad . \quad . = R_1 x - P (x - a)$

$\Delta \text{max.} \left( \text{when } a < .414l \text{ at } x = l \dfrac{l^2 + a^2}{3l^2 - a^2} \right) = \dfrac{Pa}{3EI} \dfrac{(l^2 - a^2)^3}{(3l^2 - a^2)^2}$

$\Delta \text{max.} \left( \text{when } a > .414l \text{ at } x = l \sqrt{\dfrac{a}{2l + a}} \right) = \dfrac{Pab^2}{6EI} \sqrt{\dfrac{a}{2l + a}}$

$\Delta a \quad \left( \text{at point of load} \right) . \quad . \quad . \quad . = \dfrac{Pa^2 b^3}{12EIl^3} (3l + a)$

$\Delta_x \quad \left( \text{when } x < a \right) . \quad . \quad . \quad . \quad . = \dfrac{Pb^2 x}{12EIl^3} (3al^2 - 2lx^2 - ax^2)$

$\Delta_x \quad \left( \text{when } x > a \right) . \quad . \quad . \quad . \quad . = \dfrac{Pa}{12EIl^3} (l - x)^2 (3l^2 x - a^2 x - 2a^2 l)$

## 15. BEAM FIXED AT BOTH ENDS—UNIFORMLY DISTRIBUTED LOADS

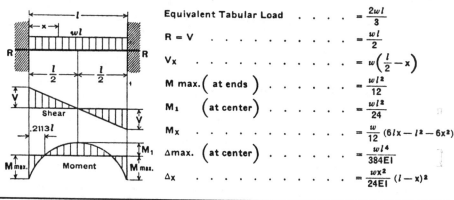

| | | |
|---|---|---|
| Equivalent Tabular Load | . . . . . | $= \dfrac{2wl}{3}$ |
| $R = V$ | . . . . . . . . . | $= \dfrac{wl}{2}$ |
| $V_x$ | . . . . . . . . . | $= w\left(\dfrac{l}{2} - x\right)$ |
| M max. $\left(\text{at ends}\right)$ | . . . . | $= \dfrac{wl^2}{12}$ |
| $M_1$ $\left(\text{at center}\right)$ | . . . . | $= \dfrac{wl^2}{24}$ |
| $M_x$ | . . . . | $= \dfrac{w}{12}(6lx - l^2 - 6x^2)$ |
| $\Delta$max. $\left(\text{at center}\right)$ | . . . | $= \dfrac{wl^4}{384EI}$ |
| $\Delta_x$ | . . . . . . . . | $= \dfrac{wx^2}{24EI}(l - x)^2$ |

## 16. BEAM FIXED AT BOTH ENDS—CONCENTRATED LOAD AT CENTER

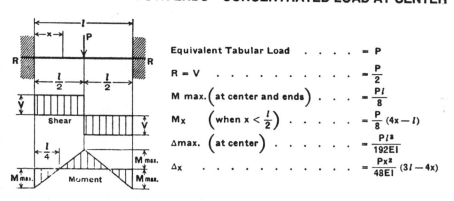

| | | |
|---|---|---|
| Equivalent Tabular Load | . . . . | $= P$ |
| $R = V$ | . . . . . . . . . | $= \dfrac{P}{2}$ |
| M max. $\left(\text{at center and ends}\right)$ | . . . | $= \dfrac{Pl}{8}$ |
| $M_x$ $\left(\text{when } x < \dfrac{l}{2}\right)$ | . . . . | $= \dfrac{P}{8}(4x - l)$ |
| $\Delta$max. $\left(\text{at center}\right)$ | . . . . | $= \dfrac{Pl^3}{192EI}$ |
| $\Delta_x$ | . . . . . . . . | $= \dfrac{Px^2}{48EI}(3l - 4x)$ |

## 17. BEAM FIXED AT BOTH ENDS—CONCENTRATED LOAD AT ANY POINT

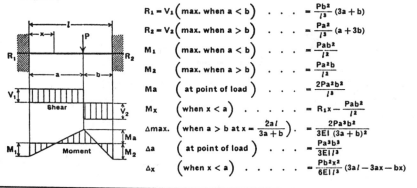

| | | |
|---|---|---|
| $R_1 = V_1 \left(\text{max. when } a < b\right)$ | . . . | $= \dfrac{Pb^2}{l^3}(3a + b)$ |
| $R_2 = V_2 \left(\text{max. when } a > b\right)$ | . . . | $= \dfrac{Pa^2}{l^3}(a + 3b)$ |
| $M_1$ $\left(\text{max. when } a < b\right)$ | . . . | $= \dfrac{Pab^2}{l^2}$ |
| $M_2$ $\left(\text{max. when } a > b\right)$ | . . . | $= \dfrac{Pa^2b}{l^2}$ |
| $M_a$ $\left(\text{at point of load}\right)$ | . . . | $= \dfrac{2Pa^2b^2}{l^3}$ |
| $M_x$ $\left(\text{when } x < a\right)$ | . . . . | $= R_1x - \dfrac{Pab^2}{l^2}$ |
| $\Delta$max. $\left(\text{when } a > b \text{ at } x = \dfrac{2al}{3a+b}\right)$ | $= \dfrac{2Pa^3b^2}{3EI(3a+b)^2}$ |
| $\Delta a$ $\left(\text{at point of load}\right)$ | . . . | $= \dfrac{Pa^3b^3}{3EIl^3}$ |
| $\Delta_x$ $\left(\text{when } x < a\right)$ | . . . . | $= \dfrac{Pb^2x^2}{6EIl^3}(3al - 3ax - bx)$ |

## 18. CANTILEVER BEAM—LOAD INCREASING UNIFORMLY TO FIXED END

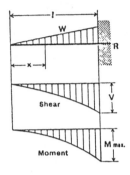

Equivalent Tabular Load . . . . $= \frac{8}{3} W$

$R = V$ . . . . . . . . . $= W$

$V_x$ . . . . . . . . . $= W \frac{x^2}{l^2}$

M max. $\left(\text{at fixed end}\right)$ . . . . . $= \frac{Wl}{3}$

$M_x$ . . . . . . . . . $= \frac{Wx^3}{3l^2}$

$\Delta$max. $\left(\text{at free end}\right)$ . . . . . $= \frac{Wl^3}{15EI}$

$\Delta_x$ . . . . . . . . . . $= \frac{W}{60EIl^2} (x^5 - 5l^4x + 4l^5)$

---

## 19. CANTILEVER BEAM—UNIFORMLY DISTRIBUTED LOAD

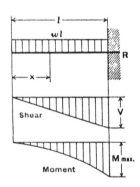

Equivalent Tabular Load . $= 4wl$

$R = V$ . . . . . . . $= wl$

$V_x$ . . . . . . . . $= wx$

M max. $\left(\text{at fixed end}\right)$ . . $= \frac{wl^2}{2}$

$M_x$ . . . . . . . . . $= \frac{wx^2}{2}$

$\Delta$max. $\left(\text{at free end}\right)$ . . $= \frac{wl^4}{8EI}$

$\Delta_x$ . . . . . . . $= \frac{w}{24EI} (x^4 - 4l^3x + 3l^4)$

---

## 20. BEAM FIXED AT ONE END, FREE BUT GUIDED AT OTHER—UNIFORMLY DISTRIBUTED LOAD

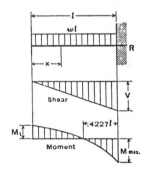

The deflection at the guided end is assumed to be in a vertical plane.

Equivalent Tabular Load . . . . $= \frac{8}{3} wl$

$R = V$ . . . . . . . . . $= wl$

$V_x$ . . . . . . . . . $= wx$

M max. $\left(\text{at fixed end}\right)$ . . . . . $= \frac{wl^2}{3}$

$M_1$ $\left(\text{at guided end}\right)$ . . . . $= \frac{wl^2}{6}$

$M_x$ . . . . . . . . . $= \frac{w}{6} (l^2 - 3x^2)$

$\Delta$max. $\left(\text{at guided end}\right)$ . . . . $= \frac{wl^4}{24EI}$

$\Delta_x$ . . . . . . . . . $= \frac{w(l^2 - x^2)^2}{24EI}$

## 1. CANTILEVER BEAM—CONCENTRATED LOAD AT ANY POINT

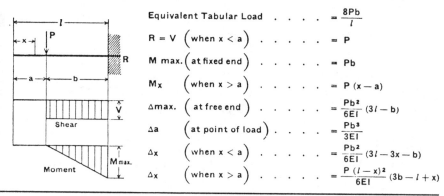

Equivalent Tabular Load $\quad = \dfrac{8Pb}{l}$

$R = V \left( \text{when } x < a \right) \quad = P$

M max. $\left( \text{at fixed end} \right) \quad = Pb$

$M_x \quad \left( \text{when } x > a \right) \quad = P(x - a)$

$\Delta$max. $\left( \text{at free end} \right) \quad = \dfrac{Pb^2}{6EI}(3l - b)$

$\Delta a \quad \left( \text{at point of load} \right) \quad = \dfrac{Pb^3}{3EI}$

$\Delta_x \quad \left( \text{when } x < a \right) \quad = \dfrac{Pb^2}{6EI}(3l - 3x - b)$

$\Delta_x \quad \left( \text{when } x > a \right) \quad = \dfrac{P(l - x)^2}{6EI}(3b - l + x)$

## 2. CANTILEVER BEAM—CONCENTRATED LOAD AT FREE END

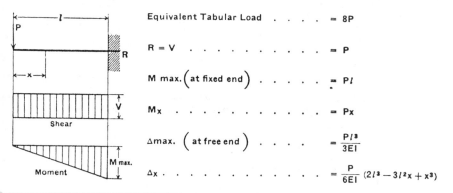

Equivalent Tabular Load $\quad = 8P$

$R = V \quad = P$

M max. $\left( \text{at fixed end} \right) \quad = Pl$

$M_x \quad = Px$

$\Delta$max. $\left( \text{at free end} \right) \quad = \dfrac{Pl^3}{3EI}$

$\Delta_x \quad = \dfrac{P}{6EI}(2l^3 - 3l^2x + x^3)$

## 3. BEAM FIXED AT ONE END, FREE BUT GUIDED AT OTHER—CONCENTRATED LOAD AT GUIDED END

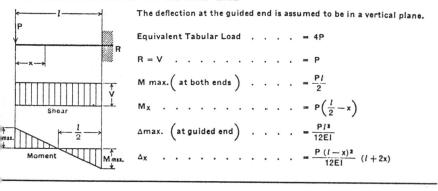

The deflection at the guided end is assumed to be in a vertical plane.

Equivalent Tabular Load $\quad = 4P$

$R = V \quad = P$

M max. $\left( \text{at both ends} \right) \quad = \dfrac{Pl}{2}$

$M_x \quad = P\left( \dfrac{l}{2} - x \right)$

$\Delta$max. $\left( \text{at guided end} \right) \quad = \dfrac{Pl^3}{12EI}$

$\Delta_x \quad = \dfrac{P(l - x)^2}{12EI}(l + 2x)$

## 24. BEAM OVERHANGING ONE SUPPORT—UNIFORMLY DISTRIBUTED LOAD

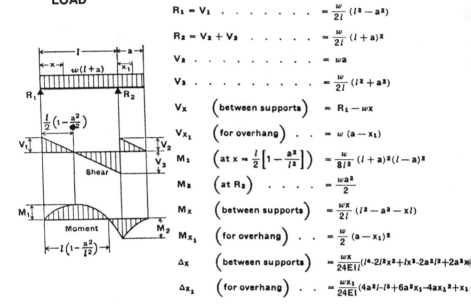

$$R_1 = V_1 \quad \ldots \ldots \ldots \quad = \frac{w}{2l}(l^2 - a^2)$$

$$R_2 = V_2 + V_3 \quad \ldots \ldots \quad = \frac{w}{2l}(l + a)^2$$

$$V_2 \quad \ldots \ldots \ldots \ldots \quad = wa$$

$$V_3 \quad \ldots \ldots \ldots \ldots \quad = \frac{w}{2l}(l^2 + a^2)$$

$$V_x \quad \left(\text{between supports}\right) \quad = R_1 - wx$$

$$V_{x_1} \quad \left(\text{for overhang}\right) \ldots \quad = w(a - x_1)$$

$$M_1 \quad \left(\text{at } x = \frac{l}{2}\left[1 - \frac{a^2}{l^2}\right]\right) \quad = \frac{w}{8l^2}(l + a)^2(l - a)^2$$

$$M_2 \quad \left(\text{at } R_2\right) \quad \ldots \ldots \quad = \frac{wa^2}{2}$$

$$M_x \quad \left(\text{between supports}\right) \quad = \frac{wx}{2l}(l^2 - a^2 - xl)$$

$$M_{x_1} \quad \left(\text{for overhang}\right) \ldots \quad = \frac{w}{2}(a - x_1)^2$$

$$\Delta_x \quad \left(\text{between supports}\right) \quad = \frac{wx}{24EIl}(l^4 - 2l^2x^2 + lx^3 - 2a^2l^2 + 2a^2x)$$

$$\Delta_{x_1} \quad \left(\text{for overhang}\right) \ldots \quad = \frac{wx_1}{24EI}(4a^2l - l^3 + 6a^2x_1 - 4ax_1^2 + x_1)$$

## 25. BEAM OVERHANGING ONE SUPPORT—UNIFORMLY DISTRIBUTED LOAD ON OVERHANG

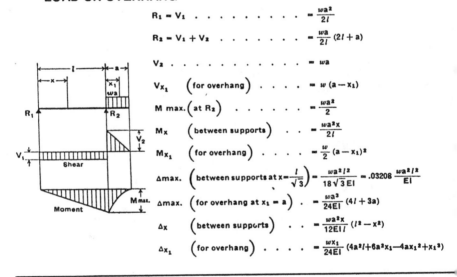

$$R_1 = V_1 \quad \ldots \ldots \ldots \ldots \quad = \frac{wa^2}{2l}$$

$$R_2 = V_1 + V_2 \quad \ldots \ldots \ldots \quad = \frac{wa}{2l}(2l + a)$$

$$V_2 \quad \ldots \ldots \ldots \ldots \ldots \quad = wa$$

$$V_{x_1} \quad \left(\text{for overhang}\right) \ldots \ldots \quad = w(a - x_1)$$

$$M \text{ max.} \left(\text{at } R_2\right) \quad \ldots \ldots \quad = \frac{wa^2}{2}$$

$$M_x \quad \left(\text{between supports}\right) \quad \ldots \quad = \frac{wa^2x}{2l}$$

$$M_{x_1} \quad \left(\text{for overhang}\right) \ldots \ldots \quad = \frac{w}{2}(a - x_1)^2$$

$$\Delta \text{max.} \quad \left(\text{between supports at } x = \frac{l}{\sqrt{3}}\right) = \frac{wa^2l^2}{18\sqrt{3}EI} = .03208\frac{wa^2l^2}{EI}$$

$$\Delta \text{max.} \quad \left(\text{for overhang at } x_1 = a\right) \quad = \frac{wa^3}{24EI}(4l + 3a)$$

$$\Delta_x \quad \left(\text{between supports}\right) \quad \ldots \quad = \frac{wa^2x}{12EIl}(l^2 - x^2)$$

$$\Delta_{x_1} \quad \left(\text{for overhang}\right) \ldots \ldots \quad = \frac{wx_1}{24EI}(4a^2l + 6a^2x_1 - 4ax_1^2 + x_1^3)$$

86

## 6. BEAM OVERHANGING ONE SUPPORT—CONCENTRATED LOAD AT END OF OVERHANG

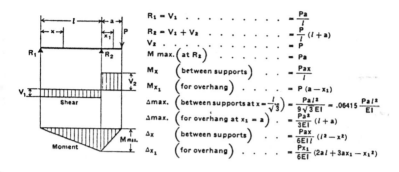

$$R_1 = V_1 \quad \ldots \ldots \ldots \ldots = \frac{Pa}{l}$$

$$R_2 = V_1 + V_2 \quad \ldots \ldots \ldots = \frac{P}{l}(l+a)$$

$$V_2 \quad \ldots \ldots \ldots \ldots \ldots = P$$

$$M \text{ max.} \left( \text{at } R_2 \right) \quad \ldots \ldots = Pa$$

$$M_x \quad \left( \text{between supports} \right) \quad \ldots = \frac{Pax}{l}$$

$$M_{x_1} \quad \left( \text{for overhang} \right) \quad \ldots = P(a - x_1)$$

$$\Delta \text{max.} \left( \text{between supports at } x = \frac{l}{\sqrt{3}} \right) = \frac{Pal^2}{9\sqrt{3}EI} = .06415 \frac{Pal^2}{EI}$$

$$\Delta \text{max.} \left( \text{for overhang at } x_1 = a \right) = \frac{Pa^2}{3EI}(l+a)$$

$$\Delta_x \quad \left( \text{between supports} \right) \quad \ldots = \frac{Pax}{6EIl}(l^2 - x^2)$$

$$\Delta_{x_1} \quad \left( \text{for overhang} \right) \quad \ldots = \frac{Px_1}{6EI}(2al + 3ax_1 - x_1^2)$$

## 7. BEAM OVERHANGING ONE SUPPORT—UNIFORMLY DISTRIBUTED LOAD BETWEEN SUPPORTS

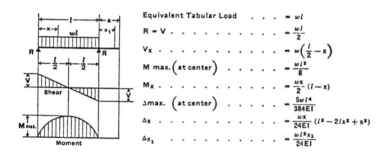

$$\text{Equivalent Tabular Load} \quad \ldots = wl$$

$$R = V \quad \ldots \ldots \ldots \ldots = \frac{wl}{2}$$

$$V_x \quad \ldots \ldots \ldots \ldots \ldots = w\left( \frac{l}{2} - x \right)$$

$$M \text{ max.} \left( \text{at center} \right) \quad \ldots = \frac{wl^2}{8}$$

$$M_x \quad \ldots \ldots \ldots \ldots \ldots = \frac{wx}{2}(l-x)$$

$$\Delta \text{max.} \left( \text{at center} \right) \quad \ldots = \frac{5wl^4}{384EI}$$

$$\Delta_x \quad \ldots \ldots \ldots \ldots \ldots = \frac{wx}{24EI}(l^3 - 2lx^2 + x^3)$$

$$\Delta_{x_1} \quad \ldots \ldots \ldots \ldots \ldots = \frac{wl^3x_1}{24EI}$$

## 8. BEAM OVERHANGING ONE SUPPORT-CONCENTRATED LOAD AT ANY POINT BETWEEN SUPPORTS

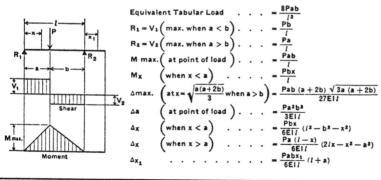

$$\text{Equivalent Tabular Load} \quad \ldots = \frac{8Pab}{l^2}$$

$$R_1 = V_1 \left( \text{max. when } a < b \right) \ldots = \frac{Pb}{l}$$

$$R_2 = V_2 \left( \text{max. when } a > b \right) \ldots = \frac{Pa}{l}$$

$$M \text{ max.} \left( \text{at point of load} \right) \ldots = \frac{Pab}{l}$$

$$M_x \quad \left( \text{when } x < a \right) \quad \ldots = \frac{Pbx}{l}$$

$$\Delta \text{max.} \left( \text{at } x = \sqrt{\frac{a(a+2b)}{3}} \text{ when } a > b \right) = \frac{Pab(a+2b)\sqrt{3a(a+2b)}}{27EIl}$$

$$\Delta_a \quad \left( \text{at point of load} \right) \quad \ldots = \frac{Pa^2b^2}{3EIl}$$

$$\Delta_x \quad \left( \text{when } x < a \right) \quad \ldots = \frac{Pbx}{6EIl}(l^2 - b^2 - x^2)$$

$$\Delta_x \quad \left( \text{when } x > a \right) \quad \ldots = \frac{Pa(l-x)}{6EIl}(2lx - x^2 - a^2)$$

$$\Delta_{x_1} \quad \ldots \ldots \ldots \ldots \ldots = \frac{Pabx_1}{6EIl}(l+a)$$

87

## 29. CONTINUOUS BEAM—TWO EQUAL SPANS—UNIFORM LOAD ON ONE SPAN

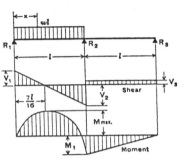

Equivalent Tabular Load . . $= \frac{49}{64} wl$

$R_1 = V_1$ . . . . . . . . $= \frac{7}{16} wl$

$R_2 = V_2 + V_3$ . . . . . $= \frac{5}{8} wl$

$R_3 = V_3$ . . . . . . . $= -\frac{1}{16} wl$

$V_2$ . . . . . . . . . $= \frac{9}{16} wl$

M Max. $\left( \text{at } x = \frac{7}{16} l \right)$ . . $= \frac{49}{512} wl^2$

$M_1$ $\left( \text{at support } R_2 \right)$ . $= \frac{1}{16} wl^2$

$M_x$ $\left( \text{when } x < l \right)$ . . . $= \frac{wx}{16} (7l - 8x)$

---

## 30. CONTINUOUS BEAM—TWO EQUAL SPANS—CONCENTRATED LOAD AT CENTER OF ONE SPAN

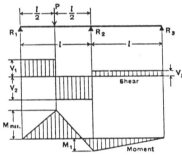

Equivalent Tabular Load . . $= \frac{13}{8} P$

$R_1 = V_1$ . . . . . . . . $= \frac{13}{32} P$

$R_2 = V_2 + V_3$ . . . . . $= \frac{11}{16} P$

$R_3 = V_3$ . . . . . . . $= -\frac{3}{32} P$

$V_2$ . . . . . . . . . $= \frac{19}{32} P$

M Max. $\left( \text{at point of load} \right)$ . $= \frac{13}{64} Pl$

$M_1$ $\left( \text{at support } R_2 \right)$ . $= \frac{3}{32} Pl$

---

## 31. CONTINUOUS BEAM—TWO EQUAL SPANS—CONCENTRATED LOAD AT ANY POINT

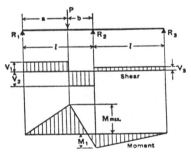

$R_1 = V_1$ . . . . . . . $= \frac{Pb}{4l^3} \left( 4l^2 - a(l+a) \right)$

$R_2 = V_2 + V_3$ . . . . . $= \frac{Pa}{2l^3} \left( 2l^2 + b(l+a) \right)$

$R_3 = V_3$ . . . . . . . $= -\frac{Pab}{4l^3} (l+a)$

$V_2$ . . . . . . . . . $= \frac{Pa}{4l^3} \left( 4l^2 + b(l+a) \right)$

M max. $\left( \text{at point of load} \right)$ . $= \frac{Pab}{4l^3} \left( 4l^2 - a(l+a) \right)$

$M_1$ $\left( \text{at support } R_2 \right)$ . $= \frac{Pab}{4l^3} (l+a)$

**FOR VARIOUS CONCENTRATED MOVING LOADS**

The values given in these formulas do not include impact which vari
according to the requirements of each case.

---

## 32. SIMPLE BEAM—ONE CONCENTRATED MOVING LOAD

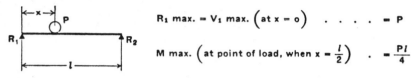

$R_1$ max. $= V_1$ max. $\left(\text{at } x = 0\right)$ . . . . . $= P$

$M$ max. $\left(\text{at point of load, when } x = \dfrac{l}{2}\right)$ . $= \dfrac{Pl}{4}$

---

## 33. SIMPLE BEAM—TWO EQUAL CONCENTRATED MOVING LOADS

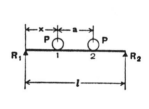

$R_1$ max. $= V_1$ max. $\left(\text{at } x = 0\right)$ . . . . $= P\left(2 - \right.$

$M$ max. $\begin{cases} \left[\begin{array}{l}\text{when } a < (2 - \sqrt{2})\ l = .586l \\ \text{under load 1 at } x = \dfrac{1}{2}\left(l - \dfrac{a}{2}\right)\end{array}\right] = \dfrac{P}{2l}\left(l - \right. \\[2em] \left[\begin{array}{l}\text{when } a > (2 - \sqrt{2})\ l = .586l \\ \text{with one load at center of span} \\ \text{(case 32)}\end{array}\right] = \dfrac{Pl}{4} \end{cases}$

---

## 34. SIMPLE BEAM—TWO EQUAL CONCENTRATED MOVING LOADS

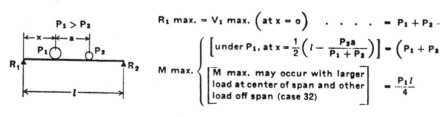

$R_1$ max. $= V_1$ max. $\left(\text{at } x = 0\right)$ . . . . . $= P_1 + P_2$

$M$ max. $\begin{cases} \left[\text{under } P_1, \text{ at } x = \dfrac{1}{2}\left(l - \dfrac{P_2 a}{P_1 + P_2}\right)\right] = \left(P_1 + P_2\right. \\[1.5em] \left[\begin{array}{l}\text{M max. may occur with larger} \\ \text{load at center of span and other} \\ \text{load off span (case 32)}\end{array}\right] = \dfrac{P_1 l}{4} \end{cases}$

---

# GENERAL RULES FOR SIMPLE BEAMS CARRYING MOVING CONCEN-TRATED LOADS

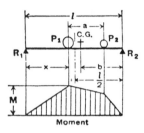

The maximum shear due to moving concentrated loads occurs at one support when one of the loads is at that support. With several moving loads, the location that will produce maximum shear must be determined by trial.

The maximum bending moment produced by moving concentrated loads occurs under one of the loads when that load is as far from one support as the center of gravity of all the moving loads on the beam is from the other support.

In the accompanying diagram, the maximum bending moment occurs under load P1 when $x = b$. It should also be noted that this condition occurs when the center line of the span is midway between the center of gravity of loads and the nearest concentrated load.

# CHAPTER 8

# DEFLECTION OF BEAMS

## 8.1   DEFLECTION AND ELASTIC CURVE

The distance traversed by a point on the neutral plane of an already straight beam in the same plane of loading is called the **deflection of the point**.

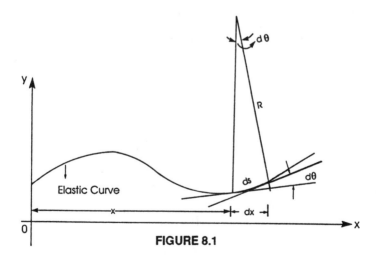

**FIGURE 8.1**

A deflection curve or an elastic curve is a line in the plane of loading collinear with the neutral axis in a bent shape.

## 8.2    DIFFERENTIAL EQUATIONS OF BEAM

In the elastic curve in Figure 8.1, consider the incremental arc $dS$ which will cause an infinitesimal change in the slope, $d\theta$, so that

$$dS = Rd\theta$$

$$\frac{d\theta}{dS} = \frac{1}{R} \tag{8-1}$$

where $R$ is the radius of curvature and $1/R$ is called curvature.

For small deflections (small angles),

$$\frac{dy}{dx} \approx \tan\theta \approx \theta, \; dS \approx dx$$

Therefore,

$$\frac{d\theta}{dS} \approx \frac{d\theta}{dx} \approx \frac{d^2y}{dx^2} \tag{8-2}$$

Comparing equations (8-1) and (8-2)

$$\frac{d^2y}{dx^2} = \frac{1}{R} \tag{8-3}$$

Now, using the flexure formula:

$$\frac{M}{I} = \frac{\sigma}{Y} = \frac{E}{R}$$

or

$$\frac{E}{R} = \frac{M}{I}$$

$$\frac{I}{R} = \frac{M}{EI} \tag{8-4}$$

Comparing equations (8-3) and (8-4)

$$\frac{M}{EI} = \frac{d^2y}{dx^2}$$

or

$$\boxed{M = EI\,\frac{d^2y}{dx^2}} \qquad (8\text{-}5)$$

This differential equation together with the boundary conditions are used to determine the deflection $y$ of the beam at any point, of distant $x$, from the origin. We can also find the shear load $Q$ by further differentiating the moment $M$, i.e.,

$$\boxed{\begin{aligned} Q &= \frac{dM}{dx} = \frac{d}{dx}\left(EI\,\frac{d^2y}{dx^2}\right) \\[2mm] \therefore\qquad Q &= EI\,\frac{d^3y}{dx^3} \end{aligned}} \qquad (8\text{-}6)$$

The above facts can be summarized as follows:

Deflection: $\Delta = y$

Slope: $\tan\theta \approx \theta \approx \dfrac{dy}{dx}$

where $\theta$ is very small.

Bending Moment: $M = EI\,\dfrac{d^2y}{dx^2}$

Curvature: $\dfrac{I}{R} = \dfrac{d^2y}{dx^2}$

Bending Moment: $M = \dfrac{EI}{R} = EI\dfrac{d^2y}{dx^2}$

Shear force: $\quad Q = \dfrac{dM}{dx} = EI\dfrac{d^3y}{dx^3}$

Load: $\quad\quad W = \int Q\,dx$

In a standard calculus text, the expression for curvature is also given as:

$$\frac{I}{R} = \frac{d^2y/dx^2}{\left[1 + \left(\dfrac{dy}{dx}\right)^2\right]^{\frac{3}{2}}} \tag{8-7}$$

The sign convention follows that the moment is positive for a beam bent downward or when the elastic curve is concave upward. Since curvature corresponds to the moment, thus a positive moment yields a positive curvature, thus giving a positive value of $\dfrac{d^2y}{dx^2}$. The slope, $\dfrac{dy}{dx}$ is also positive, corresponding to a positive moment.

# 8.3 DETERMINATION OF DEFLECTION

The deflection in the beams can be determined by the following three methods:

### 8.3.1 THE INTEGRATION METHOD

According to this method, the deflection of a beam can be determined by integrating the moment equation twice, directly from the free body diagram analysis. The constants of integrations are determined by using appropriate boundary conditions. Some interesting examples of calculating deflections in beams that are usually encountered in structural mechanics are given as follows:

## SIMPLY SUPPORTED BEAM

Considering the free body diagram in Figure 8.2(a) and writing the equation for equilibrium of moments,

$$M - \frac{wL}{2} x + \frac{wx^2}{2} = 0$$

where $M$ is the resisting moment.

Now,

$$M = \frac{wL}{2} x - \frac{wx^2}{2}$$

$$M = \frac{w}{2} x (L - x)$$

$$@ \; x = 0, \quad M = 0$$

$$x = L, \quad M = \frac{w}{2} L(L - L) = 0$$

$$x = \frac{L}{2}, \quad M = \frac{w}{2} \cdot \frac{L}{2} \left( L - \frac{L}{2} \right)$$

$$\text{or } M = \frac{wL^2}{8}$$

Thus the bending moment is zero at both ends and has a greatest value at the center with a parabolic variation along the length of the beam. (Refer to Figure 8.2(b).)

From equation (8-5),

$$M = EI \frac{d^2y}{dx^2}$$

95

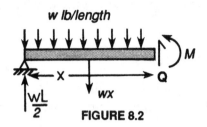

**w lb/length**

**wL/2**   **wx**   **Q**   **M**

**FIGURE 8.2**

or

$$\frac{d^2 y}{dx^2} = \frac{M}{EI} = \frac{1}{EI} \frac{w}{2} (Lx - x^2)$$

On integration,

slope,
$$\frac{dy}{dx} = \frac{1}{EI} \int \frac{w}{2} (Lx - x^2) \, dx + C_1$$

or

$$\frac{dy}{dx} = \frac{w}{2EI} \left( \frac{Lx^2}{2} - \frac{x^3}{3} \right) + C_1$$

Further integration yields,

$$y = \frac{w}{2EI} \int \left( \frac{Lx^2}{2} - \frac{x^3}{3} \right) dx + C_1 x + C_2$$

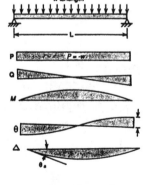

**FIGURE 8.3**

96

$$y = \frac{w}{2EI}\left(\frac{Lx^3}{6} - \frac{x^4}{12}\right) + C_1 x + C_2 \qquad (8\text{-}8)$$

The following boundary conditions are used to eliminate the constants of integration,

(i) at $x = 0$, $y = 0$

(ii) $x = L$, $y = 0$.

For the first $B.C.$, $C_2 = 0$.

For the second $B.C.$, we get,

$$0 = \frac{w}{2EI}\left[\frac{L^4}{6} - \frac{L^4}{12}\right] + C_1 L$$

or

$$0 = \frac{w}{12EI}\left[L^3 - \frac{L^3}{2}\right] + C_1$$

$$0 = \frac{w}{12EI}\left[\frac{2L^3 - L^3}{2}\right] + C_1$$

$$\therefore \quad C_1 = -\frac{wL^3}{24EI}$$

Substituting the values of $C_1$ and $C_2$ back into equation (8-8) we obtain

$$y = \frac{w}{2EI}\left(\frac{Lx^3}{6} - \frac{x^4}{12} - \frac{L^3 x}{12}\right) \qquad (8\text{-}9)$$

The deflection is maximum at $x = \frac{L}{2}$, hence

$$\Delta = \frac{w}{2EI}\left[\frac{L}{6}\left(\frac{L}{2}\right)^3 - \frac{1}{12}\left(\frac{L}{2}\right)^4 - \frac{L^3}{12}\left(\frac{L}{2}\right)\right]$$

$$= \frac{w}{2EI}\left[\frac{L^4}{48} - \frac{L^4}{48 \times 4} - \frac{L^4}{24}\right]$$

$$\therefore \quad \boxed{\Delta = -\frac{5wL^4}{384EI}} \qquad (8\text{-}10)$$

## CANTILEVER BEAM SUPPORTING A TRIANGULARLY DISTRIBUTED LOAD OF MAXIMUM INTENSITY

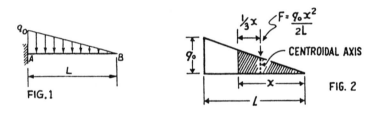

**FIGURE 8.4**

Again, as in the previous case, this problem will be solved by using the differential equation of curvature for elastic beams. To use this equation, the moment as a function of $x$ is needed.

To find the moment an imaginary section cut is made at $x$ (See Figure 8.4(b)). The internal moment acting in the beam at the cut is found by moment equilibrium. The intensity of the distributed load at $x$ is $q_0 x/L$. The average magnitude of the distributed load in the shaded section is $q_0 x/2L$. The weight of the shaded portion of the distributed loading is the average intensity times the length.

$$F = \frac{q_0 x^2}{2L}$$

The moment about point $x$ is the force times the distance to the centroidal axis of the distributed loading from $x$. For a triangle the centroidal axis is one-third of the way from the base to the apex. The moment is thus

$$\sum M_x = 0 = M - \frac{q_0 x^2}{2L} \left(\frac{x}{3}\right)$$

$$M = \frac{q_0 x^3}{6L}$$

The curvature equation can now be employed.

$$EI\frac{d^2 y}{dx^2} = M = \frac{q_0 x^3}{6L}$$

Integrating gives the slope.

$$EI\frac{dy}{dx} = \frac{q_0 x^4}{24L} + c_1$$

The integration constant can be found using a boundary condition. At the wall, the slope is zero. Therefore at $x = L$, $dy/dx = 0$. Substituting into the slope equation

$$0 = \frac{q_0 L^4}{24L} + c_1$$

Solving for $c_1$

$$c_1 = -\frac{q_0 L^3}{24}$$

The equation of the slope is thus

$$EI\frac{dy}{dx} = \frac{q_0 x^4}{24L} - \frac{q_0 L^3}{24}$$

Integrating the slope gives the deflection

$$Ely = \frac{q_0 x^5}{120L} - \frac{q_0 L^3 x}{24} + c_2$$

The integration constant, $c_2$, can also be found using boundary conditions. At the wall the deflection is zero. Therefore at $x = L$, $y = 0$. Substituting

$$0 = \frac{q_0 L^5}{120L} - \frac{q_0 L^3 (L)}{24} + c_2$$

$$c_2 = \frac{4q_0 L^4}{120} = \frac{q_0 L^4}{30}$$

The deflection equation is thus

$$Ely = \frac{q_0 x^5}{120L} - \frac{q_0 L^3 x}{24} + \frac{q_0 L^4}{30}$$

The statement of the problem asks for formulas for the slope and the deflection at the free end. At the free end $x = 0$. Substituting into the equations for slope

$$EI \frac{dy}{dx} = -\frac{q_0 L^3}{24}$$

$$\frac{dy}{dx} = -\frac{q_0 L^3}{24EI} \qquad \text{(slope)}$$

and for the deflection

$$Ely = \frac{q_0 L^4}{30}$$

$$y = \frac{q_0 L^4}{30EI} \qquad \text{(deflection)}$$

## 8.3.2    MOMENT AREA METHOD

This method is more helpful for the beams with concentrated loads, in which case discontinuities occur in the M, V, θ, and Δ diagrams. Refer to Section 9.4 for details.

Following are some illustrations of the application of the moment-area method.

### 8.3.2.1    DEFLECTION AND SLOPE OF A SIMPLY-SUPPORTED BEAM DUE TO THE CONCENTRATED LOAD P

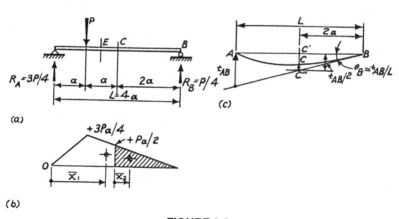

**FIGURE 8.5**

The bending-moment diagram is in Figure 8.5(a). Since $EI$ is constant, the $M/(EI)$ diagram need not be made, as the areas of the bending-moment diagram divided by $EI$ give the necessary quantities for use in the moment-area theorems. The elastic curve is in Figure 8.5(c). It is concave upward throughout its length as the bending moments are positive. This curve must pass through the points of the support at $A$ and $B$.

101

It is apparent from the sketch of the elastic curve that the desired quantity is represented by the distance $CC'$. Moreover, from purely geometrical or kinematic considerations, $CC' = C'C'' - C''C$, where the distance $C''C$ is measured from a tangent to the elastic curve passing through the point of support $B$. However, since the deviation of a support point from a tangent to the elastic curve at the other support may always be computed by the second moment-area theorem, a distance such as $C'C''$ may be found by proportion from the geometry of the figure. In this case, $t_{AB}$ follows by taking the whole $M/(EI)$ area between $A$ and $B$ and multiplying it by its $\bar{x}$ measured from a vertical through $A$, whence $C'C'' = \frac{1}{2} t_{AB}$. By another application of the second theorem, $t_{CB}$, which is equal to $C''C$, is determined. For this case, the $M/(EI)$ area is shaded in Figure 8.5(b), and, for it, the $\bar{x}$ is measured from $C$. Since the right reaction is $P/4$ and the distance $CB = 2a$, the maximum ordinate for the shaded triangle is $+Pa/2$.

$$v_C = C'C'' - C''C = (t_{AB}/2) - t_{CB}$$

$$t_{AB} = \Phi_1 \bar{x}_1 = \frac{1}{EI}\left(\frac{4a}{2}\frac{3Pa}{4}\right)\frac{(a+4a)}{3} = +\frac{5Pa^3}{2EI}$$

$$t_{CB} = \Phi_2 \bar{x}_2 = \frac{1}{EI}\left(\frac{2a}{2}\frac{Pa}{2}\right)\frac{(2a)}{3} = +\frac{Pa^3}{3EI}$$

$$v_C = \frac{t_{AB}}{2} - t_{CB} = \frac{5Pa^3}{4EI} - \frac{Pa^3}{3EI} = \frac{11Pa^3}{12EI}$$

The positive signs of $t_{AB}$ and $t_{CB}$ indicate that points $A$ and $C$ lie above the tangent through $B$. As may be seen from Figure 8.5(a), the deflection at the center of the beam is in a downward direction.

The slope of the elastic curve at $C$ can be found from the slope at one of the ends. For point $B$ on the right

$$\theta_B = \theta_C + \Delta\theta_{BC} \text{ or } \theta_C = \theta_B - \Delta\theta_{BC}$$

$$\theta_C = \frac{t_{AB}}{L} - \Phi_2 = \frac{5Pa^2}{8EI} - \frac{Pa^2}{2EI} = \frac{Pa^2}{8EI}$$

radians counterclockwise.

## 8.3.2.2    ANGLES OF ROTATION

Angles of rotation, $\theta_a$ and $\theta_b$, at the ends of the beam and the deflection $\delta$ at the middle of a simple beam acted upon by couples $M_0$ at the ends:

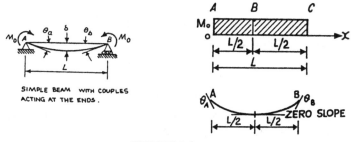

SIMPLE BEAM WITH COUPLES ACTING AT THE ENDS.

**FIGURE 8.6**

**SOLUTION:** This problem will be solved using the first and second moment area theorems.

The moment diagram is given in Figure 8.6(b) . The slope at the mid-point of the beam, $x = L/2$, is zero (See Figure 8.6(c).

By the first moment-area theorem, the difference in slope between points $A$ and $B$ is

$$\theta_B - \theta_A = \int_A^B \frac{M}{EI} \, dx$$

Since $\theta_B$ is zero and the value of the integral is the area of the moment diagram (Figure 8.6(b)) between $A$ and $B$ ($M_0L/2$) di-

103

vided by $EI$ the slope at $A$ is

$$\theta_A = -\frac{M_0 L}{2EI}$$

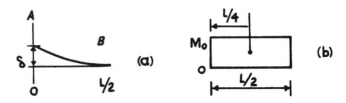

**FIGURE 8.7**

Again by the first moment-area theorem the difference in slope between points $B$ and $C$ is

$$\theta_C - \theta_B = \int_B^C \frac{M}{EI}\, dx$$

The area of the moment diagram $Mx$ from $B$ to $C$ is $M_0 L/2$ and $\theta_B$ is zero. Substituting yields

$$\theta_C = \frac{M_0 L}{2EI}$$

By the second moment-area theorem, the deflection is from Figure 8.7.

$$\delta = \frac{A\bar{x}}{EI} = \frac{M_0\left(\dfrac{L}{2}\right)\left(\dfrac{L}{4}\right)}{EI} = \frac{M_0 L^2}{8EI}$$

The deflection at the middle of the beam is $M_0 L^2/8EI$ downwards.

The results obtained can be checked by referring to a table of beam deflections and slopes. If the case of a simply supported

104

beam with couples acting at both ends is not listed, the formulas for a beam with a couple acting at only one end can be modified using superposition. This results in

$$-\theta_a = \theta_b = \frac{M_0 L}{3EI} + \frac{M_0 L}{6EI} = \frac{M_0 L}{2EI}$$

$$\delta = (2)\frac{M_0 L^2}{16EI} = \frac{M_0 L^2}{8EI}$$

This agrees with the results obtained using the moment-area theorems.

### 8.3.2.3    SIMPLY SUPPORTED BEAM

A simply supported beam having maximum deflection and rotation of the elastic curve at the ends caused by the application of a uniformly distributed load of $P_0$ lb per foot, Figure 8.8(a). Assume constant $EI$.

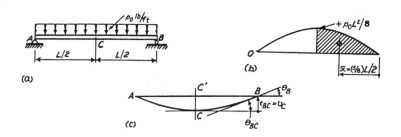

**FIGURE 8.8**

The bending-moment diagram is in Figure 8.8(a). It is a second-degree parabola with a maximum value at the vertex of $P_0 L^2/8$. The elastic curve passing through the points of the support $A$ and $B$ is shown in Figure 8.8(a).

In this case, the $M/(EI)$ diagram is symmetrical about a

105

vertical line passing through the center. Therefore the elastic curve must be symmetrical, and the tangent to this curve at the center of the beam is horizontal. From the figure, it is seen that $\Delta\theta_{BC}$ is equal to $\theta_B$, and the rotation of the end $B$ is equal to one-half the area of the whole $M/(EI)$ diagram. The distance $CC'$ is the desired deflection, and from the geometry of the figure it is seen to be equal to $t_{BC}$.

$$\Phi = \frac{1}{EI}\left(\frac{2}{3}\frac{L}{2}\frac{P_0L^2}{8}\right) = \frac{P_0L^3}{24\,EI}$$

$$\theta_B = \Delta\theta_{BC} = \Phi = +\frac{P_0L^3}{24\,EI}$$

$$v_C = V_{max} = t_{BC} = \Phi\bar{x} = \frac{P_0L^3}{24\,EI}\frac{5L}{16} = \frac{5P_0L^4}{384\,EI}$$

Since the point $B$ is above the tangent through $C$, the sign of $v_C$ is positive.

### 8.3.2.4 ANGLE OF ROTATION $\theta_b$ AND DEFLECTION $\delta$ AT THE FREE END OF A CANTILEVER WITH A CONCENTRATED LOAD P (SEE FIGURE 8.9)

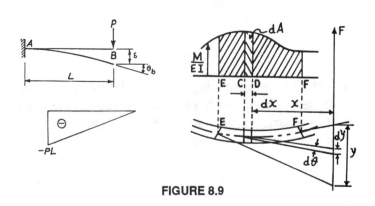

**FIGURE 8.9**

106

This problem can be solved using the first and second moment area theorems (bending moment diagram is triangular in shape and is shown in the lower part of the figure).

## SECOND MOMENT-AREA THEOREM

On any vertical line, the distance $y$, between the intersections made by the tangents at points $E$ and $F$ is equal to the first moment about that line of the area of the $M/(EI)$ graph between the points $E$ and $F$.

From the first moment-area theorem we observe that the difference in angles between points $A$ and $B$ is equal to the area of the bending moment diagram divided by $EI$; this is $-PL^2/2EI$. The minus sign means that the tangent at $B$ is rotated clockwise from the tangent at $A$, which is horizontal. Therefore, the angle $\theta_b$, positive as shown in the figure, is

$$\theta_b = \frac{PL^2}{2EI}$$

The deflection $\delta$ at the end of the beam can be obtained by applying the second theorem. The distance $\Delta$ of point $B$ on the deflection curve from the tangent at $A$ is equal to the first moment of the bending moment area about $B$ divided by $EI$:

$$\Delta = -\frac{PL^2}{2EI}\left(\frac{2L}{3}\right) = -\frac{PL^3}{3EI}$$

The minus sign means that point $B$ on the deflection curve is below the tangent at $A$. The deflection $\delta$, therefore is

$$\delta = \frac{PL^3}{3EI}$$

## 8.3.2.5 DEFLECTION OF THE FREE END OF THE CANTILE-VER BEAM OF FIGURE 8.10(a) IN TERMS OF w, L, E AND I

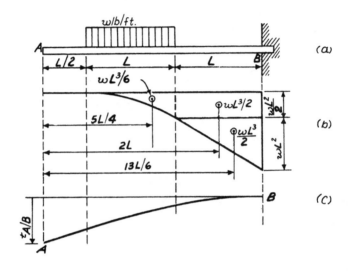

**FIGURE 8.10**

Since $E$ and $I$ are constant, a bending moment diagram is used instead of an $M/(EI)$ diagram. The moment diagram is shown in Figure 8.10(b), and the area under it has been divided into rectangular, triangular, and parabolic parts for ease in calculating areas and moments. The elastic curve is shown in Figure 8.10(c) with the deflection greatly exaggerated. Points A and B are selected at the ends of the beam because the beam has a horizontal tangent at $B$ and the deflection at $A$ is required. The vertical distance to A from the tangent at $B$, $t_{A/B}$ in Figure 8.10(c), equals the deflection of the free end of the beam, $y_A$. For this reason, the second area-moment theorem can be used directly to obtain the required deflection. The area of each of the three portions of the area under the moment diagram is shown in

108

Figure 8.10(b) along with the distance from the centroid of each part to the moment axis at the free end $A$. The second area-moment theorem gives

$$EIt_{A/B} = -\frac{wL^3}{6}\left(\frac{5L}{4}\right) - \frac{wL^3}{2}(2L) - \frac{wL^3}{2}\left(\frac{13L}{6}\right),$$

from which

$$y_A = t_{A/B} = -\frac{55wL^4}{24EI} = \frac{55wL^4}{24EI}$$

downward.

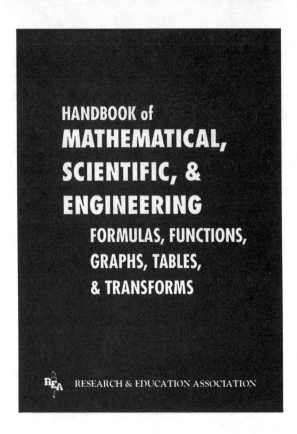

**HANDBOOK of MATHEMATICAL, SCIENTIFIC, & ENGINEERING**

**FORMULAS, FUNCTIONS, GRAPHS, TABLES, & TRANSFORMS**

RESEARCH & EDUCATION ASSOCIATION

A particularly useful reference for those in math, science, engineering and other technical fields. Includes the most-often used formulas, tables, transforms, functions, and graphs which are needed as tools in solving problems. The entire field of special functions is also covered. A large amount of scientific data which is often of interest to scientists and engineers has been included.

*Available at your local bookstore or order directly from us by sending in coupon below.*

*Available at your local bookstore or order directly from us by sending in coupon below.*

# HANDBOOK AND GUIDE FOR
# COMPARING and SELECTING
# COMPUTER LANGUAGES

| | |
|---|---|
| BASIC | PL/1 |
| FORTRAN | APL |
| PASCAL | ALGOL-60 |
| COBOL | C |

- This book is the first of its kind ever produced in computer science.
- It examines and highlights the differences and similarities among the eight most widely used computer languages.
- A practical guide for selecting the most appropriate programming language for any given task.
- Sample programs in all eight languages are written and compared side-by-side. Their merits are analyzed and evaluated.
- Comprehensive glossary of computer terms.

*Available at your local bookstore or order directly from us by sending in coupon below.*

# "The ESSENTIALS"
# of Math & Science

Each book in the ESSENTIALS series offers all essential information of the field it covers. It summarizes what every textbook in the particular field must include, and is designed to help students in preparing for exams and doing homework. The ESSENTIALS are excellent supplements to any class text.

The ESSENTIALS are complete, concise, with quick access to needed information, and provide a handy reference source at all times. The ESSENTIALS are prepared with REA's customary concern for high professional quality and student needs.

## Available in the following titles:

| | | |
|---|---|---|
| Advanced Calculus I & II | Electronic Communications I & II | Modern Algebra |
| Algebra & Trigonometry I & II | Electronics I & II | Numerical Analysis I & II |
| Anthropology | Finite & Discrete Math | Organic Chemistry I & II |
| Automatic Control Systems / | Fluid Mechanics / | Physical Chemistry I & II |
| Robotics I & II | Dynamics I & II | Physics I & II |
| Biology I & II | Fourier Analysis | Set Theory |
| Boolean Algebra | Geometry I & II | Statistics I & II |
| Calculus I, II & III | Group Theory I & II | Strength of Materials & |
| Chemistry | Heat Transfer I & II | Mechanics of Solids I & II |
| Complex Variables I & II | LaPlace Transforms | Thermodynamics I & II |
| Differential Equations I & II | Linear Algebra | Topology |
| Electric Circuits I & II | Math for Engineers I & II | Transport Phenomena I & II |
| Electromagnetics I & II | Mechanics I, II & III | Vector Analysis |

*If you would like more information about any of these books,
complete the coupon below and return it to us or go to your local bookstore.*

## RESEARCH & EDUCATION ASSOCIATION
61 Ethel Road W. • Piscataway, New Jersey 08854
Phone: (908) 819-8880

### Please send me more information about your Essentials Books

Name _____

Address _____

City _____ State _____ Zip _____

# REA's Test Preps
# The Best in Test Preparations

The REA "Test Preps" are far more comprehensive than any other test series. They contain more tests with much more extensive explanations than others on the market. Each book provides several complete practice exams, based on the most recent tests given in the particular field. Every type of question likely to be given on the exams is included. Each individual test is followed by a complete answer key. **The answers are accompanied by full and detailed explanations.** By studying each test and the pertinent explanations, students will become well-prepared for the actual exam.

REA *has published 40 Test Preparation volumes in several series. They include:*

**Advanced Placement Exams (APs)**
Biology
Calculus AB & Calculus BC
Chemistry
Computer Science
English Literature & Composition
European History
Government & Politics
Physics
Psychology
United States History

**College Board Achievement Tests (CBATs)**
American History
Biology
Chemistry
English Composition

French
German
Literature
Mathematics Level I, II & IIC
Physics
Spanish

**Graduate Record Exams (GREs)**
Biology
Chemistry
Computer Science
Economics
Engineering
General
History
Literature in English
Mathematics
Physics
Political Science
Psychology

**CBEST** - California Basic Educational Skills Test
**CDL** - Commercial Drivers License Exam
**ExCET** - Exam for Certification Educators in Texas
**FE (EIT)** - Fundamentals of Engineering Exam
**GED** - High School Equivalency Diploma Exam
**GMAT** - Graduate Management Admission Test
**LSAT** - Law School Admission
**MCAT** - Medical College Admission Test
**NTE** - National Teachers Exam
**SAT** - Scholastic Aptitude Test
**TOEFL** - Test of English as a Foreign Language

---

**RESEARCH & EDUCATION ASSOCIATION**
61 Ethel Road W. • Piscataway, New Jersey 08854
Phone: (908) 819-8880

**Please send me more information about your Test Prep Books**

Name _____

Address _____

City _____ State _____ Zip _____

# REA's Problem Solvers

The "PROBLEM SOLVERS" are comprehensive supplemental text-books designed to save time in finding solutions to problems. Each "PROBLEM SOLVER" is the first of its kind ever produced in its field. It is the product of a massive effort to illustrate almost any imaginable problem in exceptional depth, detail, and clarity. Each problem is worked out in detail with step-by-step solution, and the problems are arranged in order of complexity from elementary to advanced. Each book is fully indexed for locating problems rapidly.

ADVANCED CALCULUS
ALGEBRA & TRIGONOMETRY
AUTOMATIC CONTROL
   SYSTEMS/ROBOTICS
BIOLOGY
BUSINESS, MANAGEMENT, & FINANCE
CALCULUS
CHEMISTRY
COMPLEX VARIABLES
COMPUTER SCIENCE
DIFFERENTIAL EQUATIONS
ECONOMICS
ELECTRICAL MACHINES
ELECTRIC CIRCUITS
ELECTROMAGNETICS
ELECTRONIC COMMUNICATIONS
ELECTRONICS
FINITE & DISCRETE MATH
FLUID MECHANICS/DYNAMICS
GENETICS
GEOMETRY

HEAT TRANSFER
LINEAR ALGEBRA
MACHINE DESIGN
MATHEMATICS for ENGINEERS
MECHANICS
NUMERICAL ANALYSIS
OPERATIONS RESEARCH
OPTICS
ORGANIC CHEMISTRY
PHYSICAL CHEMISTRY
PHYSICS
PRE-CALCULUS
PSYCHOLOGY
STATISTICS
STRENGTH OF MATERIALS &
   MECHANICS OF SOLIDS
TECHNICAL DESIGN GRAPHICS
THERMODYNAMICS
TOPOLOGY
TRANSPORT PHENOMENA
VECTOR ANALYSIS

*If you would like more information about any of these books,*
*complete the coupon below and return it to us or go to your local bookstore.*

---

**RESEARCH & EDUCATION ASSOCIATION**
61 Ethel Road W. • Piscataway, New Jersey 08854
Phone: (908) 819-8880

**Please send me more information about your Problem Solver Books**

Name _____

Address _____

City _____ State _____ Zip _____